Taking Flutter to the Web

Learn how to build cross-platform UIs for web and mobile platforms using Flutter for Web

Damodar Lohani

BIRMINGHAM—MUMBAI

Taking Flutter to the Web

Group Product Manager: Pavan Ramchandani

Publishing Product Manager: Aaron Tanna

Senior Editor: Sofi Rogers

Senior Content Development Editor: Rakhi Patel

Technical Editor: Saurabh Kadave

Copy Editor: Safis Editing

Language Support Editors: Sofi Rogers and Safis Editing

Project Coordinator: Manthan Patel

Proofreader: Safis Editing

Indexer: Manju Arsan

Production Designer: Shyam Sundar Korumilli

Marketing Coordinator: Anamika Singh

First published: October 2022

Production reference: 1280922

Published by Packt Publishing Ltd.
Livery Place
35 Livery Street
Birmingham
B3 2PB, UK.

ISBN 978-1-80181-771-4

www.packt.com

To my father, astrologer Pandit Ojaraj Upadhyaya Lohani, and my mother, Gita Lohani, for their sacrifices, support, and encouragement. To my wife, Pooja, for being a loving partner, and to my children, for their unfathomable love and respect.

– Damodar Lohani

Contributors

About the author

Damodar Lohani is a software engineer at Appwrite. He is also a Google Developer Expert for Flutter and Dart. He has been working in the industry as a full stack developer specializing in different technologies since 2012 for diverse clients. He also runs the Flutter Kathmandu (`https://meetup.com/flutter-kathmandu`) community and organizes Flutter meetups in Kathmandu to help Flutter enthusiasts come together and help each other out. He is also a speaker and mentor; he has spoken at events such as Flutter Vikings and Flutter Yatra and mentored at numerous hackathons and events in Nepal. He is also actively involved in open source building and sharing his own open source projects as well as contributing to other open source projects. You can find his open source projects at `https://github.com/lohanidamodar`.

In 2015, he graduated with a BSc in computer science and information technology from Tribhuvan University, Kathmandu, Nepal. He has been using Flutter to build mobile and web applications for over 3 years.

I want to thank all the people who have been directly and indirectly responsible for making this book a success. I want to thank my wife Pooja for her unquestionable support during my late nights and early endeavors to finish the content of this book in time. I want to thank my parents for their unrestricted love and support. I want to thank my children, for they have sacrificed their personal time with me to let me work in peace to complete this book.

I want to thank Eldad Faux, founder of Appwrite, and my teammates at Appwrite for always supporting and motivating me to complete this book. I want to thank all my friends and relatives who are directly and indirectly involved in getting me here in life.

About the reviewer

Adby Santos has a degree in information technology from UFERSA. He is a co-organizer and contributor in Flutterando, one of the world's biggest communities about Flutter, and he now works at FTeam, a mobile apps specialist organization, as a Flutter development specialist. He is also a speaker and consultant who has spoken at and moderated events such as Flutter Global Summit by Geekle. He also writes weekly about his professional and personal experiences on his LinkedIn.

Table of Contents

3

Building Responsive and Adaptive Designs 31

Part 2: Flutter Web under the Hood

4

Flutter Web under the Hood 49

5

Understanding Routes and Navigation 57

6

Architecting and Organizing   71

Part 3: Advanced Concepts

7

Implementing Persistence   93

8

State Management in Flutter   111

11

Preface

Taking Flutter to the Web is a book about Flutter and Dart that teaches you to take your basic Flutter knowledge and build responsive web applications.

Who this book is for

This book is intended for Flutter mobile developers and Dart programmers who want to consolidate their basic Flutter knowledge and build responsive web applications with Flutter.

What this book covers

Chapter 1, *Getting Started with Flutter on the Web*, introduces the Flutter ecosystem and the web as a part of that ecosystem. It also makes clear why we should learn Flutter on the web and what benefits it can add. Finally, it describes what types of web apps should and shouldn't be built with Flutter. You will also get to know some real-world Flutter web applications.

Chapter 2, *Creating Your First Web App*, will describe how to start a Flutter project with web platform support. We will build and run a Flutter project on the web using Flutter widgets. This will be a dynamic and responsive app that interacts with its audience.

Chapter 3, *Building Responsive and Adaptive Designs*, will teach us that the web has a huge number of target devices of different screen sizes and densities. As Flutter is cross-platform, each platform needs to have a native feel. This chapter covers the concept of adaptive and responsive design, which lets us just build apps that are responsive and adaptive to the platform – the web.

Chapter 4, *Flutter Web under the Hood*, will look at how Flutter renders its widgets as web apps. Understanding these inner workings will help us build better web apps using Flutter.

Chapter 5, *Understanding Routes and Navigation*, will detail how to use routes and navigation in Flutter on the web. This chapter will cover navigation with a special focus on how this occurs on the web. You will see how to use declarative navigation and understand its benefits.

Chapter 6, *Architecting and Organizing*, will cover the need for proper architecture and we will learn about simple yet scalable architecture. This chapter discusses why scalable architecture is important and how you can organize files and folders to make it easier to scale and work. You will understand the MVVM architecture, which is simple yet scalable, and see how to practically apply it in an application.

Chapter 7, Implementing Persistence, will teach us how to save data locally using various options – simple key-value storage, or more advanced object storage. You will discover how to persist simple key-value data, which is useful for user settings, and how to use the offline storage for caching, as well as key-value data using HiveDB.

Chapter 8, State Management in Flutter, will outline the basics of state management, as well as covering how to use Riverpod as a state management solution. You will discover the importance of state management and the various options available for it.

Chapter 9, Integrating Appwrite, will teach us how to integrate Appwrite, an open source BaaS for web and mobile applications with services such as Authentication, Database, Cloud Functions, and more.

Chapter 10, Firebase Integration, will detail how to integrate and use Firebase solutions to build dynamic web apps using Flutter. We will cover the authentication part, wherein we will be able to authenticate users. We will see how to implement Firestore to save and load dynamic content.

Chapter 11, Building and Deploying a Flutter Web Application, will expand upon the concepts required to build and deploy web apps developed with Flutter. In this final chapter, we will also learn how to automate our task of building and deploy web apps using CI/CD.

To get the most out of this book

In order to get the most out of this book, you need to have some basic knowledge and understanding of programming.

We assume you have basic knowledge and understanding of the following:

- Basic programming
- Working with the command line
- The Dart programming language
- The Flutter framework

To gain the most benefits, it is always advised to practice by following along with the lessons. When the lesson is code-heavy, it is advised to keep calm and proceed one step at a time.

Software/hardware covered in the book	Operating system requirements
Flutter 3	Windows, macOS, or Linux
Dart 2.17	

To follow along and run the codes provided for each chapter, you need to install Flutter. You can do so by following the official guide at `https://docs.flutter.dev/get-started/install`. To build and run Flutter and Dart on the web once they are installed, all you need is to install the Google Chrome browser.

If you are using the digital version of this book, we advise you to type the code yourself or access the code from the book's GitHub repository (a link is available in the next section). Doing so will help you avoid any potential errors related to the copying and pasting of code.

Download the example code files

You can download the example code files for this book from GitHub at `https://github.com/PacktPublishing/Taking-Flutter-to-the-Web`. If there's an update to the code, it will be updated in the GitHub repository.

We also have other code bundles from our rich catalog of books and videos available at `https://github.com/PacktPublishing/`. Check them out!

Download the color images

We also provide a PDF file that has color images of the screenshots and diagrams used in this book. You can download it here: `https://packt.link/YvdST`

Conventions used

There are a number of text conventions used throughout this book.

`Code in text`: Indicates code words in the text, database table names, folder names, filenames, file extensions, pathnames, dummy URLs, user input, and Twitter handles. Here is an example: "Then, in the `main.dart` file, first import `firebase_core` and `firebase_options.dart` generated by `flutterfire_cli`."

A block of code is set as follows:

```
Future<bool> anonymousLogin() async {
  if (isLoggedIn) {
    error = 'Already logged in';
    return false;
  }
}
```

Any command-line input or output is written as follows:

```
flutter pub get
```

Bold: Indicates a new term, an important word, or words that you see onscreen. For instance, words in menus or dialog boxes appear in **bold**. Here is an example: "Click on the **Start new project** button."

> **Tips or important notes**
> Appear like this.

Get in touch

Feedback from our readers is always welcome.

General feedback: If you have questions about any aspect of this book, email us at customercare@packtpub.com and mention the book title in the subject of your message.

Errata: Although we have taken every care to ensure the accuracy of our content, mistakes do happen. If you have found a mistake in this book, we would be grateful if you would report this to us. Please visit www.packtpub.com/support/errata and fill in the form.

Piracy: If you come across any illegal copies of our works in any form on the internet, we would be grateful if you would provide us with the location address or website name. Please contact us at copyright@packt.com with a link to the material.

If you are interested in becoming an author: If there is a topic that you have expertise in and you are interested in either writing or contributing to a book, please visit authors.packtpub.com.

Share Your Thoughts

Once you've read *Taking Flutter to the Web*, we'd love to hear your thoughts! Scan the QR code below to go straight to the Amazon review page for this book and share your feedback.

https://packt.link/r/1801817715

Your review is important to us and the tech community and will help us make sure we're delivering excellent quality content.

Part 1: Basics of Flutter Web

In this part of the book, you will be introduced to the web platform and the importance of learning Flutter web.

This part comprises the following chapters:

- *Chapter 1, Getting Started with Flutter on the Web*
- *Chapter 2, Creating Your First Web App*
- *Chapter 3, Building Responsive and Adaptive Design*

1
Getting Started with Flutter on the Web

Flutter recently has become a very popular choice for building cross-platform applications. Flutter has gained a lot of popularity among mobile application developers and start-ups. With the introduction of stable web support and the preview of support for desktop, Flutter now supports building on six different platforms. In this chapter, we will learn how we can get started with Flutter web development.

By the end of this chapter, you will have a good understanding of the basics of Flutter on the web. You will understand what Flutter on the web is and how to get started with it. You will also understand how Flutter on the web is different from regular web and mobile development. The knowledge you gain in this chapter will also help you understand what you can build with Flutter on the web and when to use it.

In this chapter, we will cover the following main topics:

- Introduction to Flutter and Flutter on the web
- Why you should learn Flutter on the web
- What you can build with Flutter on the web
- Making and running a new Flutter web project
- Flutter on the web for web developers
- Flutter on the web for mobile developers
- Official Flutter documentation

Technical requirements

The technical requirements for this chapter are as follows:

- A computer with a decent spec that can run Flutter. You can read the official Flutter getting started documentation for the required spec. Visit `https://flutter.dev/docs/get-started/install` and select your operating system to see the system requirements.

- Flutter installed and running with web support enabled.

You can download the latest code samples for this chapter from the book's official GitHub repository at `https://github.com/PacktPublishing/Taking-Flutter-to-the-Web/tree/main/Chapter01/chapter1_final`.

Introduction to Flutter and Flutter on the web

Imagine being able to write code in one language/framework and deploy it on six different platforms, with some of those platforms having completely different setups in terms of both UI and UX. This has been made possible by Flutter in a way that allows developers to respect the norms of each platform, enriching developers with the tools and techniques to develop on each of these platforms with a native feel. A cross-platform development framework has been needed by developers for a long time now, because of the high costs and length of development time for native code on each platform. Many frameworks have tried, but none of them have succeeded in the way that Flutter has. Flutter has worked openly in the community in close connection with developers to resolve their issues and problems. Focusing both on the developer experience as well as the native feel and performance of applications, Flutter has become a great tool for cross-platform application development.

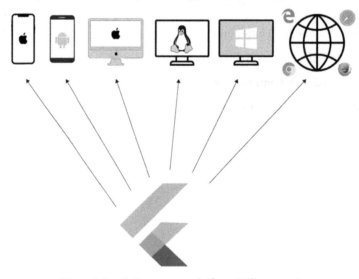

Figure 1.1 – Flutter, a cross-platform UI framework

Imagine developing a mobile application for Android and iOS, and imagine writing the same code without any changes (or with just a few tweaks) and running on the web while maintaining the same amount of functionality, or more. Imagine that same code deploying native desktop apps for Mac, Windows, and Linux. Yes, that's now possible with Flutter.

From Flutter version 2.0, Flutter web support has become stable. This means we are now able to build Android, iOS, and web apps from a single code base. Flutter's web support delivers the same experiences on the web as on mobile. Using the same sets of widgets that you use to build a UI for mobile applications, you can now build and deploy web applications. This enables developers to build feature-rich, interactive, and graphics-intense web applications with Flutter.

However, you have to understand that as Flutter is an app-centric UI framework, it is different from traditional web development. We are not writing HTML, CSS, and **JavaScript** (**JS**). Instead, we're developing web apps using the same widgets that we use to develop mobile applications with Flutter. Later on in this chapter, we will see how Flutter is different from traditional web development with HTML, CSS, and JS, and how to take advantage of those differences.

Flutter is a lucrative framework to learn. With its plugins and developer tooling ecosystem, the Flutter framework has become one of the best tools for cross-platform application development in recent years.

Why you should learn Flutter on the web

If you are already developing mobile applications with Flutter, imagine the same code base deploying to the web with only a few or no tweaks required. Imagine not even having to write native code most of the time. If you don't know much about the web, it doesn't matter – with Flutter, you don't need to. However, I do suggest gaining basic knowledge of the web as a platform and how it's different from mobile applications. We will discuss this topic later in this chapter.

If you are a developer looking to enhance your toolbox, improve your career, and you are interested in mobile development, Flutter is the right choice for 2022. Not only is Flutter popular among developer communities, but it also has lots of job opportunities in the industry as Flutter has become one of the best cross-platform frameworks of choice among companies as well as developers.

It's easy to learn Flutter as there are lots of resources online. There are tons of free and paid courses, articles, and videos. Thanks to Flutter being a popular choice in the community and being open source, many developers have created a lot of free content around it. There are thousands of open source example projects on GitHub, from small examples to full-fledged real-world applications. Having this amount of references makes it easy to learn. As Flutter is so popular among the developer community, many problems you may have while learning Flutter will have already been solved by the community. You can easily find answers on platforms such as GitHub and Stack Overflow.

Those of you who are confused about which framework to learn, as there are many alternatives, such as Kotlin Multiplatform, .NET MAUI, React Native, Ionic, Apache Cordova, and more that I may not even know about, I implore you to try Flutter at least once. I myself have tried both React Native and Flutter, and later have completely switched to Flutter as I love the whole ecosystem of tools it provides.

What you can build with Flutter on the web

We can build any kind of app in Flutter and then successfully port it to the web. Does that mean we should build all web apps with Flutter? If not, how do we decide what and what not to build with Flutter on the web? The following scenarios are some examples of when it would be beneficial to choose Flutter instead of traditional forms to build web applications. You can also refer to these great case studies of applications built with Flutter on the web available on the official Flutter documentation while deciding what kind of applications benefit from Flutter on the web:

- Supernova (`https://flutter.dev/showcase/supernova`)
- Rive (`https://flutter.dev/showcase/rive`)
- iRobot (`https://flutter.dev/showcase/irobot`)
- Flutter Gallery (`https://gallery.flutter.dev/`)

And here are some community examples, built with Flutter on the web:

- Flutterando Community (`https://flutterando.com.br/#/`)
- Flutter Shape Maker (`https://shapemaker.web.app/#/`)
- Rows Spreadsheet App (`https://rows.com/`)

Progressive Web Applications

Progressive Web Applications (**PWAs**) are a quite recently introduced form of web application that tries to bring the best of both worlds – applications and the web. We will learn more about PWAs later, in *Chapter 11, Building and Deploying a Flutter Web App*. Flutter delivers high-quality PWAs that are integrated with a user's environment, including installation, offline support, and tailored UX that is the same as Flutter's mobile application. So, at the moment, Flutter is a very suitable choice for building a PWA.

Single-page applications

With Flutter being a client-side framework, it is a great option for building complex standalone web apps that are rich with graphics and interactive content to reach end users on a wide variety of devices.

Existing mobile applications

If you already have a mobile application built with Flutter and want to target new users on a web platform, Flutter on the web would be a great choice. With few or no modifications to your existing code, you will be able to easily port your existing application to the web and target a wide range of users and devices.

Now, before we dive into developing awesome web apps using Flutter, read and keep in mind the following paragraph from the official Flutter documentation (`https://flutter.dev/web`):

> *Not every HTML scenario is ideally suited for Flutter at this time. For example, text-rich, flow-based, static content such as blog articles benefit from the document-centric model that the web is built around, rather than the app-centric services that a UI framework like Flutter can deliver. However, you can use Flutter to embed interactive experiences into these websites.*

Here, we need to understand that Flutter is an app-centric UI framework designed to build interactive experiences, which may not be suitable for all kinds of websites and web applications.

Deciding when to use Flutter on the web

This is not a definitive guide to help you decide when to use Flutter on the web. However, this might help you in the process of deciding whether or not to build your next web app with Flutter.

With web applications, one thing I can say for certain is that if you have the time and budget to build a native web application, build it. Build a native web application with HTML, CSS, and JS, the languages of the web. Also, if the web is the primary focus of your business, and most of your target customers are on the web, I suggest going native.

However, in scenarios in which you have mobile applications built with Flutter and most of your customers are on mobile applications but you want to quickly try the web platform to get more target customers and get more traction for your business, Flutter on the web would be a quick and cost-effective choice.

When you are planning to build a PWA or a graphics-intense, interactive application, Flutter would still be a suitable choice. Choosing Flutter will also allow you to build for mobile and desktop platforms using the same code base.

Making and running a new Flutterweb project

We'll assume that you have already installed Flutter and the IDE of your choice. If you have not already, you can follow the official installation guide (`https://flutter.dev`) to install Flutter. You also need to install the Chrome browser in order to develop and run Flutter web apps.

In order to create a Flutter project with web support, make sure you are using Flutter 2.0 or newer. It's best to use the latest stable release of Flutter, which at the time of writing this book is 3.0.5. In order to verify that you have installed Flutter and configured it to work properly, you can use the `doctor` command. In your terminal, enter the following command:

```
flutter doctor
```

If everything is set up correctly, you should see a message similar to the following:

```
[√] Flutter (Channel stable, 3.0.5, on Linux, locale en_US.UTF-
8)
[√] Android toolchain - develop for Android devices (Android
SDK version 30.0.3)
[√] Chrome - develop for the web
[√] Linux toolchain - develop for Linux desktop
[√] Android Studio (version 4.2)
[√] IntelliJ IDEA Community Edition (version 2021.1)
[√] Connected device (2 available)
[√] HTTP Host Availability
```

Based on the platform you are on and the tools you have set up, yours might have a few differences, but you should see Chrome checked, Flutter checked, version 2.0 or newer, and Visual Studio Code or Android Studio checked as the IDE.

If Chrome is not enabled and this is the first time you are using Flutter, then you may need to enable Flutter on the web. For that, you can use the following command in the terminal:

```
flutter config --enable-web
```

If you are not using the latest version of Flutter, you can do so by running the upgrade command from the terminal:

```
flutter upgrade
```

If Flutter is set up correctly, you will now be able to create your first project. In order to create your project, from the terminal window, go to your project folder and run the following command to create a new project:

```
flutter create flutter_blog
```

This command will create a new Flutter project called flutter_blog in your project folder. You can now use your favorite IDE to open the project. We will be using Visual Studio Code with Flutter and Dart extensions installed. The directory structure of your project should look like the following:

```
flutter_blog
├── android
├── ios
├── lib
├── web
└── pubspec.yaml
```

Make sure you have a web folder. If not, you will have to check your Flutter installation again and make sure everything is correct and that the web platform is enabled. If it's not enabled, enable the web platform as discussed in the previous steps, and create a new project again.

Now, in order to run the project on the web, run the following command from your terminal:

```
flutter run -d chrome
```

If everything goes well, you should see the following output in the web browser:

Figure 1.2 – Flutter Demo Home Page

Now, if we inspect the web app, we can see that Flutter widgets don't compile to traditional HTML elements. We would rather see a canvas element that is rendering our whole app. To understand how this renders in the HTML page, you will have to look at web/index.html. Let's look at it. The complete source code for the file can be found at https://github.com/PacktPublishing/ Taking-Flutter-to-the-Web/blob/main/Chapter01/chapter1_final/web/ index.html.

There is lots of HTML code. The important part is this script, which loads the Flutter Engine, which in turn loads our built app's JavaScript code:

```
...
    window.addEventListener('load', function(ev) {
        // Download main.dart.js
        _flutter.loader.loadEntrypoint({
            serviceWorker: {
```

```
          serviceWorkerVersion: serviceWorkerVersion,
        }
      }).then(function(engineInitializer) {
        return engineInitializer.initializeEngine();
      }).then(function(appRunner) {
        return appRunner.runApp();
      });
    });
  ...
```

Apart from the HTML boilerplate code, the only important block of code is this script in the body of the HTML. This script loads the dart file that is generated by Flutter and also does some other jobs to check whether Flutter Engine is initialized and make server workers work for PWAs. Flutter generates a single dart file for our dart code, and that script is what renders our Flutter application.

If you look at lib/main.dart, there is no difference in the code. The same code that builds the mobile app also builds the web app.

This is the main function that begins execution while building or running this application. This is the entry point for our Flutter application and loads the root widget that handles the rendering of our application UI:

```
void main() {
  runApp(MyApp());
}
```

It is a combination of the root widget and other widgets that renders the UI of the application:

```
class MyApp extends StatelessWidget {
  @override
  Widget build(BuildContext context) {
    return MaterialApp(
      title: 'Flutter Demo',
      theme: ThemeData(
        primarySwatch: Colors.blue,
      ),
      home: MyHomePage(title: 'Flutter Demo Home Page'),
    );
  }
}
```

This is a simple stateful widget that, when built, displays basic text with a counter. It provides a floating action button, which, when tapped, increases the counter:

```
class MyHomePage extends StatefulWidget {
  MyHomePage({Key? key, required this.title}) : super(key:
    key);
  final String title;
  @override
  _MyHomePageState createState() => _MyHomePageState();
}

class _MyHomePageState extends State<MyHomePage> {
  int _counter = 0;
  void _incrementCounter() {
    setState(() {
      _counter++;
    });
  }

  @override
  Widget build(BuildContext context) {
    return Scaffold(
      appBar: AppBar(
        title: Text(widget.title),
      ),
      body: Center(
        child: Column(
          mainAxisAlignment: MainAxisAlignment.center,
          children: <Widget>[
            Text(
              'You have pushed the button this many
              times:',
            ),
            Text(
              '$_counter',
              style: Theme.of(context).textTheme.headline4,
            ),
```

```
      ],
    ),
  ),
  floatingActionButton: FloatingActionButton(
    onPressed: _incrementCounter,
    tooltip: 'Increment',
    child: Icon(Icons.add),
  ),
);
}
}
```

So, there is not actually any difference in the code we write for mobile and web for a simple application such as this. However, as the application grows and starts to have multiple features, we will have to decide how each feature looks and works for different platforms, and embed platform-specific logic. That is when the code will start to look different, but it will still be the same code that runs on mobile as well as the web.

Once you run or build a Flutter application for the web, in your project folder there will be a build folder, inside which there is a web folder. Now if you look inside the build folder, you have similar files as you have in the project's root web folder. The `index.html` file is the exact same file. But there's a new `main.dart.js` file that is loaded in `index.html`. `main.dart.js` is the file that contains all the code that we have written in Flutter. All the code that we have written is compiled and minified into one JavaScript file and that is what loads our app's logic and UI.

Flutter on the web for web developers

If you come from a web development background, this section is for you. Even if you don't come from a development background, this section still might help you in understanding Flutter on the web in a more meaningful way. You cannot look at Flutter on the web the same way you look at web development. If you look at it that way, you will find lots of things about Flutter on the web that you will not like. Like some others online, you may feel that Flutter on the web is not performant, not ready for the web yet, and so on. As we have already talked about what types of apps benefit from Flutter on the web and what types of apps are not yet suitable to build using Flutter on the web earlier in this chapter, there's no point arguing again here about the performance and web readiness of Flutter.

To understand Flutter web development, first, you need to understand how Flutter works for mobile apps and how it's unique.

Flutter as a UI framework

You need to look at Flutter on the web from the fresh perspective that it's fundamentally different from web development using HTML. Flutter is designed to develop rich, interactive UIs, whereas the web traditionally was designed to serve textual content. But with advances in web technologies, nowadays we deliver all sorts of content on the web, including images, videos, and even graphics intensive games. You also need to think from the point of view that, even though the web is a different platform, Flutter on the web works a bit differently. Flutter is an app-centric UI framework. The web apps built with Flutter provide more of an app-like experience, instead of a traditional web-like experience. We should aim to use that to our advantage.

The first difference we need to understand is that the UI and logic we write with Flutter don't translate to HTML elements, CSS, and JS one to one. Almost every piece of logic and UI that we write is translated to JavaScript, and the UI is rendered using a combination of the Canvas element, some CSS, SVGs, WebGL, and Web Assembly. Flutter has two different renderers for the web: an HTML renderer, and a CanvasKit renderer. We will learn in detail about these in *Chapter 4, Web Flutter under the Hood*. However, the fundamental difference is that the HTML renderer uses combinations of HTML elements, CSS, Canvas elements, and SVG elements, and is smaller in terms of download size. Whereas, CanvasKit is fully consistent with Flutter mobile and desktop and has faster performance with higher widget density, but has a bigger download size.

The next difference is that Flutter is not designed for a traditional web experience. We should also keep in mind that the web has also changed a lot since it was originally launched as a medium for sharing text. Only because of this advancement in the web has it been possible to serve all sorts of content including videos, games, and other rich and interactive UIs. Applications built with Flutter are suitable for the modern web.

The next thing that you need to understand is that Flutter on the web is a single-page, client-side application. When a user requests the application, the whole application is loaded entirely in memory. The application requests the information required for different parts of the application and loads it dynamically during execution. That also means that server-side rendering is not yet possible for Flutter web applications. Everything must work on the client. Because of this, SEO is terrible at the moment with Flutter. Though there are some packages in the works by the community, such as https://pub. dev/packages/seo_renderer, which tries to resolve the SEO problem, there hasn't been any significant progress. So, applications such as blogs might not yet be suitable to be built using Flutter.

Finally, you need to understand that Flutter is not designed to build web pages or websites as we know them. We are not building Flutter web apps to serve traditional web content. Flutter is used to build app-centric, graphics-rich, and interactive applications. Flutter web applications are able to provide the same experience as their mobile application counterparts.

So, while thinking about building your next Flutter web application, do not think in terms of traditional web pages and websites. Rather, think of a mobile application or highly interactive content being delivered via the web.

Flutter on the web for mobile developers

This section is for you if you come from a mobile development background and try to approach Flutter on the web in the same way you would mobile development. You must understand one important thing: the web is fundamentally a different platform than mobile. It has tons of differences compared to the mobile platform.

First and foremost, unlike mobile applications that target specific platforms and devices with specific capabilities, a web application might run on a wide range of devices including mobile devices, desktops, laptops, and even embedded devices. Each of these devices has a different **operating system (OS)** with different capabilities under the hood. Accessing device functionalities is not as easy as with mobile devices as there are many security concerns as well as performance issues. Though with HTML5 there has been a lot of improvement in accessing device capabilities, it is still not as powerful as mobile APIs themselves, due to the wide range of devices available for web browsing. As web developers, we will have to think of all those responsibilities. Though most of the work is already delegated to plugins with Flutter, we still must think of using those plugins effectively and also think of situations where certain functionalities may not be available and the app will still have to be usable.

The next thing to remember is that the mobile application is downloaded and installed on the user's device. This means the application lives on the user's device, whereas a web application lives somewhere in the network and is only downloaded once the user requests to view it via their browser. In most cases, when a web application is viewed, only the requested page or part of the application is downloaded to the user's device. However, as we already discussed in the previous section, Flutter's web application is a single-page web application, so the whole application is loaded when the user requests the application. In many cases, when the user wants to use the application again, they will have to download the application again, making a request via their browser. The key thing to understand from this difference is that a web application has to think carefully about its size, as it's downloaded every time a user wants to use that application. In recent years, PWAs have come into being, and are essentially trying to be the best of both worlds. We will talk in detail about PWAs in later chapters.

Another thing to notice here is that as the mobile app lives on a user's device, the user will have to download updates each time the developer pushes new updates. The user may decide not to update, however, when it comes to web apps living on the server, once the developer updates them, users will get the updated version on their next request. So, a web app is a lot easier to maintain compared to mobile apps.

Unlike mobile applications, a web application also has to take into account a wide range of devices and platforms. Each device and platform has its own pros and cons. The web developer has to think of each platform and device, too. There is a huge variety of possible devices, each with different screen sizes and densities, different platforms, and OSs with different capabilities that the web app could run on. There are also concepts such as **Responsive** and **Adaptive** designs, where the design should be made dynamic to adapt to each platform and device. As a web developer, you will have to think of how to leverage the tools provided by the framework to create a web application that feels and behaves natively on any platform it is accessed on, no matter the capabilities of that platform. We will discuss more on this topic in *Chapter 3, Building Responsive and Adaptive Designs*.

Therefore, when you look at using Flutter for building web apps, you have to think beyond mobile development. In a web application, you will have to think of a wide range of screen sizes and the capabilities of devices that might run the application. You'll need to think of the wide possibilities of different platforms. You'll also have to think of the initial load time, as all the resources will have to be downloaded with the request.

Official Flutter documentation

You can find the official Flutter documentation at `https://flutter.dev/docs`. It's the best place to look for getting started with tutorials, references, and official guides. It also has great resources for those coming from other platforms such as React Native, Android, iOS, the web, or Xamarin.Forms. We will try to centralize web-specific knowledge in this book, but always remember to go back to the docs when you want to learn more and proceed beyond the scope of this book.

Summary

In this chapter, we introduced you to Flutter on the web and also explained why you should learn Flutter. We then learned what it is good to build with Flutter on the web and what is not suitable (at this current time) to build with Flutter on the web. We also built and ran a default Flutter starter project on the web platform. Finally, we also introduced Flutter on the web from the perspective of a web developer and a mobile developer, and we described how building for Flutter on the web can be different from regular web app development and regular mobile development.

In the next chapter, we will start diving deep into creating our project for the book. We will begin by building a basic layout for our application.

2

Creating Your First Web App

In the previous chapter, we introduced you to Flutter on the web. In this chapter, we will start exploring the basics of Flutter web. We will begin by creating a new Flutter project, which we will complete during the course of this book.

By the end of this chapter, you will be able to create and run a new Flutter project with web support. You will also learn about basic Flutter widgets, and then go on to learn about Flutter layout widgets to build different layouts. We will also develop a basic UI required for our application. We will be building an online learning platform, a lightweight version of Udemy.

In this chapter, we will cover the following main topics:

- Creating a new Flutter project with Flutter web
- Using basic widgets
- Building layouts

Technical requirements

The technical requirements for this chapter are as follows:

- Flutter version 3.0 or later installed and running
- Visual Studio Code or Android Studio
- Google Chrome browser

You can download the code samples for this chapter and other chapters from the book's official GitHub repository at `https://github.com/PacktPublishing/Taking-Flutter-to-the-Web`. The starter code for this chapter can be found inside the `chapter2_start` folder.

Creating a new Flutter project with Flutter web

In the last chapter, we already created our Flutter project with web support. In this chapter, we will start our project for this book. As before, let's create our new project:

```
flutter create flutter_academy
```

This command will create a new Flutter project, and if you have followed the steps from the previous chapter, you should already have web enabled. This new project should be created with web support.

Now, make sure you can run your project using the following command from your terminal, or by tapping *F5* after opening the project in Visual Studio Code:

```
flutter run -d chrome
```

As we will have already coded some starter code, copy the files inside `chapter2_start/lib` to the `lib` folder of the project you just created. We have some basic widgets already set up here, which we will talk about in the next section.

Using basic widgets

We will start by revising some basics, which will involve creating sections of our home page. To begin, open the `chapter2_start/lib/widgets/featured_section.dart` file and start updating the code as the following.

First, we will create a stateless widget:

```dart
import 'package:flutter/material.dart';

class FeaturedSection extends StatelessWidget {
  const FeaturedSection({ Key? key }) : super(key: key);

  @override
  Widget build(BuildContext context) {
    return Container();
  }
}
```

Here, we are creating a stateless widget that just displays a blank container. Our featured section will have an image, a title, a description, and a button. We want to display the image on the left and content on the right, or vice versa. Let's add our image and content. Update the body of the `build` method, as follows:

```
return Container(
  child: Row(
    children: [
      Image.asset("image"),
      Text("Title"),
      Text("Description"),
      ElevatedButton(
        child: Text("Button"),
        onPressed: (){},
      )
    ],
  ),
);
```

Here, we added a Row widget as a child of the container. The Row widget allows us to render its child widget horizontally. However, we want the content to be vertically stacked one after another. So, let's update the children of row, as follows:

```
children: [
  Expanded(child: Image.asset("image")),
  Expanded(
    child: Column(
      children: [
        Text("Title"),
        Text("Description"),
        ElevatedButton(
          child: Text("Button"),
          onPressed: () {},
        )
      ],
    ),
  )
],
```

Here, we are wrapping our Image widget with an Expanded widget. That will expand to take the available space on the row. We are also wrapping the right-side content with an Expanded widget, so both contents will take exactly 50% of the available space. In order to display the content vertically, we are using the Column widget. Now, all that remains to do is to add parameters to receive the image path, title, description, button text, and the button onPressed handler. Let's add those parameters:

```
class FeaturedSection extends StatelessWidget {
  const FeaturedSection({
    Key? Key,
    required this.image,
    required this.title,
    required this.description,
    required this.buttonLabel,
    required this.onActionPressed,
  }) : super(key: key);

  final String image;
  final String title;
  final String description;
  final String buttonLabel;
  final Function() onActionPressed;
...
```

We have updated the class constructor, as well as added class properties to accept the title, description, button label, and button-pressed function handler. Now, we will use these in our widgets. To do that, update the build method with the following:

```
return Container(
  child: Row(
    children: [
      Expanded(child: Image.asset(image)),
      Expanded(
        child: Column(
          children: [
            Text(title),
            Text(description),
            ElevatedButton(
              child: Text(buttonLabel),
              onPressed: onActionPressed,
            )
          ],
        ),
      ),
    )
```

```
      ],
    ),
  );
```

Here, we are using the properties to display our image and content instead of hardcoding the values. This will allow us to use this widget multiple times, each instance with different values. Using the FeaturedSection widget we have created here will enable multiple featured sections with different properties to appear on our home page. Finally, we will now add some styles to our code.

Let's first style the title. Replace the title text with the following:

```
Text(
  title,
  style: Theme.of(context).textTheme.headline3,
),
```

Here, we are giving the headline3 text theme for the title. The style comes from the default theme applied to our application, so that later we can update the theme to update the overall style of the text everywhere in the app, instead of updating individual styles.

Finally, we will add some spacing and alignments to make the widget look nicer. Update the build method, as follows:

```
return Container(
  width: 1340,
  padding: const EdgeInsets.all(32.0),
  child: Row(
    children: [
      Expanded(
        child: Image.asset(
          image,
          height: 450,
        ),
      ),
      const SizedBox(width: 20.0),
      Expanded(
        child: Column(
          children: [
            Text(
              title,
```

```
            style: Theme.of(context).textTheme.headline3,
          ),
          const SizedBox(height: 20.0),
          Text(description),
          const SizedBox(height: 10.0),
          ElevatedButton(
            child: Text(buttonLabel),
            onPressed: onActionPressed,
          )
        ],
      ),
    )
  ],
 ),
);
```

Here, we have added some spacing between contents using the SizedBox widget. We have also given a fixed height to our Image to keep the size constant, no matter the size of the image used. We also gave a fixed width to our wrapping container so that it won't take up the whole screen's width. Then, we added some padding around the content by passing padding to the wrapping container. This completes our featured section widget. The complete code for this can be found in the chapter2_final/lib/widgets/featured_section.dart folder.

Let's also build the CourseCard widget that we will use to display the list of courses on our home page. We will begin by creating a new stateless widget and then add some properties, as we did in the previous section. Create a new file called chapter2/lib/widgets/course_card.dart and update it as follows:

```
import 'package:flutter/material.dart';

class CourseCard extends StatelessWidget {
  const CourseCard({
    Key? key,
    required this.image,
    required this.title,
    required this.onActionPressed,
    required this.description,
  }) : super(key: key);
```

```
    final String image;
    final String title;
    final Function() onActionPressed;
    final String description;

    @override
    Widget build(BuildContext context) {
        return Container();
    }
}
```

Here, we created a simple stateless widget and added properties for the image, title, description, and action handler, which is similar to what we used before to build our FeaturedSection widget. Now, we will update the build method, as follows, to build the layout:

```
return Container(
    width: 350.0,
    child: Card(
        child: InkWell(
            onTap: onActionPressed,
            child: Column(
                children: [
                    Image.asset(
                        image,
                        height: 250,
                    ),
                    const SizedBox(height: 10.0),
                    Text(
                        title,
                        style: Theme.of(context).textTheme.headline4,
                    ),
                    const SizedBox(height: 10.0),
                    Text(description)
                ],
            ),
        ),
    ),
);
```

Similar to what we have done before, we are building this layout using a `Card` widget in combination with the `Column` and `InkWell` widgets to make the whole card clickable. We have also given some fixed sizes as well as some stylings.

So, we've now covered the basics of widgets. We are not going too deep into this, as you should have already covered these concepts while learning the basics of Flutter. You can view other widgets that we have already built for you in the `chapter2/lib/widgets` folder. They too are widgets that we will use in the next section to build the layout of our home page. All the widgets in the `widgets` folder are built using basic widgets similar to what we have looked at in this chapter.

> **Tip**
>
> If you have any queries, revisit the basics of Flutter and widgets in *Chapter 1, Getting Started with Flutter on the Web.*

Building layouts

Let's continue by building our home page, and in the process, we will learn about some more basic layout widgets that will be essential in further chapters when we're learning about more advanced concepts. We will be reusing the widgets we built in the last section as well as the ones already provided with the starter project in the `chapter2/lib/widgets` folder. Let us start by creating our home page. Create a new file called `chapter2/lib/pages/home_page.dart`, and start by creating a stateless widget with the following:

```dart
import 'package:flutter/material.dart';

class HomePage extends StatelessWidget {
  @override
  Widget build(BuildContext context) {
    return Scaffold(
      body: ListView(
        padding: const EdgeInsets.all(16.0),
        children: <Widget>[
          AppBar(
            title: Text('Flutter Academy'),
          ),
        ],
      ),
    );
  }
}
```

Here, we are introducing new widgets that will help build layouts. Firstly, we will be using the Material Design widgets. For each page, we will use the `Scaffold` widget, which provides the basic layout for a page. Then, we have `ListView`, which will allow us to render a list of widgets, such as `Row` or `Column` that we used in the previous section. However, unlike rows and columns, the `ListView` widget allows us to render content that can overflow the screen. The `ListView` widget automatically handles the overflowing content by providing a scrolling container. In our `ListView`, we have only added an `AppBar` widget for now, which will be our top navigation. We will now build our home page by using the different widgets provided to us in the `chapter2/lib/widgets` folder.

First, let's add a header section below `AppBar`. For that, import the `header.dart` file, which contains the header widget at the top:

```
import 'package:flutter_academy/widgets/header.dart';
```

Once you import the header, you can now use the widget, so below `AppBar`, add the `Header` widget with the following code:

```
...
Header(),
...
```

After adding the header, our home page is starting to take shape. Now, you can run the app on the web using the following command from your terminal:

```
flutter run -d web
```

Once you run this, your app should build and run on a Chrome browser, and you should see the following output:

Figure 2.1 – Home page with header

The second section will be a list of recent courses. For this, we will add a horizontally scrollable list of courses in a card view. If you look inside the chapter2/lib/widgets folder, there is a file named course_card.dart that contains our course card view. We will use this and build our recent courses section. Now, below the Header widget, we will add a horizontally scrolling ListView using the following code:

```
const SizedBox(height: 40.0),
Padding(
  padding: const EdgeInsets.only(left: 20.0),
    child: Text("Recent Courses",
      style: Theme.of(context).textTheme.headline3),
),
const SizedBox(height: 10.0),
Container(
  height: 450,
  child: ListView(
    scrollDirection: Axis.horizontal,
    children: [
    // Add course cards
    ],
  ),
),
```

Here, we start by adding a 40-point gap, followed by a heading that says **Recent Courses**. Then, we add a container and give it a fixed height; otherwise, ListView inside another ListView would cause an overflow error. Giving this fixed height makes it possible to render the horizontal ListView. It also ensures that all of our course cards will have the same fixed height.

Finally, inside ListView, we will add our CourseCard. As we did before, we will first import the course card, as follows:

```
import 'package:flutter_academy/widgets/course_card.dart';
```

Now, we add the course cards as the children of ListView by adding the following:

```
...
children: [
  const SizedBox(width: 20.0),
  CourseCard(
    title: "Taking Flutter to Web",
    image: Assets.course,
```

```
    description:
        "Flutter web is stable. But there are no proper
        courses focused on Flutter web. So, in this course
        we will learn what Flutter web is good for and we
        will build a production grade application along
        the way.",
    onActionPressed: () {},
  ),
  const SizedBox(width: 20.0),
  CourseCard(
    title: "Flutter for Everyone",
    image: Assets.course,
    description:
        "Flutter beginners' course for everyone. For those
        who know basic programming, can easily start
        developing Flutter apps after taking this
        course.",
    onActionPressed: () {},
  ),
  // ... you can add more courses
],
...
```

Here, we added two `CourseCard` widgets separated by a width of 20 points. You can add more cards in the same way. If you now hit **hot restart**, you should see the following output:

Figure 2.2 – Home page with course cards

Now, we have our header and our recent courses. This means we are ready to add some featured sections to display different information. We will add two featured sections below the container showing the list of courses. Make sure you import the `FeaturedSection` widget as we imported other widgets before:

```
...
Center(
  child: FeaturedSection(
    image: Assets.instructor,
    title: "Start teaching today",
    description:
        "Instructors from around the world teach millions
        of students on Udemy. We provide the tools and
        skills to teach what you love.",
    buttonLabel: "Become an instructor",
    onActionPressed: () {},
  ),
),
Center(
  child: FeaturedSection(
    imageLeft: false,
    image: Assets.instructor,
    title: "Transform your life through education",
    description:
        "Education changes your life beyond your
        imagination. Education enables you to achieve your
        dreams.",
    buttonLabel: "Start learning",
    onActionPressed: () {},
  ),
),
...
```

The following screenshot shows how the two featured sections look. We are adding two variations, where one has the image on the left and the other has the image on the right:

Figure 2.3 – Home page featured section

Now it's time to complete our layout by adding a call to action section, then one more featured section, and finally a footer. We will do that by following the same process as we have up until now. Firstly, we will import our call to action and footer widgets, as we have already imported the featured section widget:

```
import 'package:flutter_academy/widgets/call_to_action.dart';
import 'package:flutter_academy/widgets/footer.dart';
```

Now, let's add these sections below the featured section we've already added, by using the following code:

```
...
CallToAction(),
Center(
  child: FeaturedSection(
    imageLeft: false,
    image: Assets.instructor,
    title: "Know your instructors",
    description:
        "Know your instructors. We have chosen the best of
        them to give you highest quality courses.",
    buttonLabel: "Browse",
    onActionPressed: () {},
  ),
),
Footer(),
...
```

Here, we have added the call to action widget, a featured section widget as we did previously, and finally, a footer widget that contains various links and some more information about our product.

This screenshot shows the sections we just added:

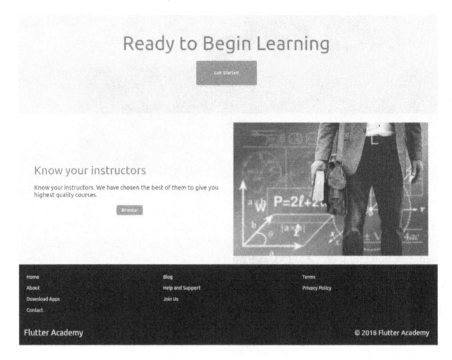

Figure 2.4 – Call to action and footer

Well done! We have successfully built the layout of our home page widget! We've also revised the basics of widgets.

Summary

In this chapter, we revised the basics of widgets. We built simple widgets, as well as a complex home page layout using a variety of the widgets we built. Now that we have revised the basics of widgets, we can move on to more advanced topics in the next chapter. We will start the next chapter by learning about responsive and adaptive design. We will also modify our home page to be responsive, to fit both large and small screen sizes.

3

Building Responsive and Adaptive Designs

In the previous chapter, we learned about the basics of Flutter and built a basic home page layout for our Flutter Academy application. In this chapter, we will learn concepts such as responsive and adaptive design and update our home page using these concepts.

By the end of this chapter, you will understand what responsive and adaptive design is. You will also understand why you need to make your apps responsive and adaptive and why it is even more important in Flutter. Along with this, you will gain the necessary skills to make your app responsive and adaptive. Finally, you will also learn how to make the UI of your Flutter Academy application responsive and adaptive for the web.

In this chapter, we will cover the following main topics:

- What is responsive and adaptive design?
- Why there's a need for responsive and adaptive design in Flutter
- Responsive and adaptive design tools you may have known and used
- Tools and techniques available in Flutter to make designs responsive and adaptive
- Making our app responsive and adaptive

Technical requirements

The technical requirements for this chapter are as follows:

- Flutter version 3.0 or later installed and running
- Visual Studio Code or Android Studio
- A Google Chrome browser

You can download the code samples of this chapter and other chapters from the book's official GitHub repository at `https://github.com/PacktPublishing/Taking-Flutter-to-the-Web`. The starter code for this chapter can be found inside the `chapter3_start` folder.

What is responsive and adaptive design?

First, let's understand the difference between responsive and adaptive applications. They are separate dimensions of an application. An app can be either responsive or adaptive, both, or none. In this section, I will give you detailed information about the difference between these two. So, let's get started.

Responsive design

A responsive app is one whose layout is tuned for multiple screen sizes that the app is targeted for. The layout adapts to the changing screen size no matter what the target device is. This often means that if the user resizes the window or the orientation changes, then we can relay the UI to adapt to the change in the size of the viewport. This is especially necessary for web applications, as they can run on all sorts of devices, each with a screen of a different size and density.

Adaptive design

An adaptive app is one that runs on different types of devices and adapts to the norms of each device type. This is a bit different from the adaptive designs that you may have heard about in a web design context. For example, on mobile, an app has to deal with touch input, whereas on desktop, it has to deal with mouse and keyboard input. Each platform has different expectations about the app's visuals. If we want to make users feel at home, we need to make our applications adapt to the expectations of each device type. This also involves using platform-specific features. For a framework such as Flutter, this also takes care of the differences in each platform.

Now that we know what responsive and adaptive designs are, let's see why we need to consider these approaches while designing our application in the next section.

Why there's a need for responsive and adaptive design in Flutter

Now you know about what responsive and adaptive design is. So, as a Flutter developer, why should you care about responsive and adaptive design? Flutter enables you to build apps that can run on mobile, desktop, and the web from a single code base. However, this raises new challenges. Though you are using the same code base, you want your app to feel familiar to users, which means adapting to each platform. So, as well as being multiplatform, the app should also be fully platform-adaptive.

Another challenge is the range of device sizes the app supports. Sizes range from mobile devices to desktops and anything that can run web applications. Thus, you need to build an app that is responsive to the screen sizes the app gets loaded into. This also means making the app responsive to the changing screen size and layouts.

Another challenge with a cross-platform framework is the different forms of input. For example, mobile devices have touch input, whereas desktops have mouse and keyboard input. An application while running on mobile should consider the touch input and make tappable items larger so that it's easier to tap with a finger. The same application running on desktops can have smaller tappable items, as the mouse will be the primary source of input. Similarly, unlike mobile devices, you will have to think of keyboard shortcuts and keyboard inputs while running on desktop devices.

With a framework such as Flutter, it's very necessary to implement the concepts of responsive and adaptive design to give an application a proper user experience, no matter what the platform is and what screen size the application is being loaded into. By implementing responsive and adaptive design, users of each platform will feel right at home with the application. An application made with Flutter for multiple platforms from a single code base, if properly implemented, will allow users to feel like it has been built natively for their platform.

Now that we understand the need for considering responsive and adaptive approaches while developing our application, let's see what responsive and adaptive design tools you may already know and use in the next section.

Responsive and adaptive design tools you may have known and used

If you already are a developer, you might have come across the concept of responsive design and might even have used some existing tools on different platforms to build responsive applications. If you are a web developer, you must be familiar with **Cascading Style Sheets** (**CSS**) media queries. CSS media queries are used on the web to handle different screen sizes to make application layouts. By using CSS media queries, you can define different CSS styling for your application based on the screen size. The screen size is provided, and media queries are handled by the browser.

If you are an Android developer, then you might have used another technique to handle different screen sizes of Android devices to make different layouts. Android allows you to provide different layout files, each in a specific folder, for device orientation and screen sizes. Here, you write a totally different layout file for each screen size and orientation that you want to handle. Any size or orientation that you don't have a custom layout for will use the default layout.

Here, we learned about different responsive and adaptive design tools that you may have already used. In the next section, we will learn about the tools and techniques available in Flutter for responsive and adaptive design.

Tools and techniques available for responsive and adaptive design

Flutter provides different tools and techniques for developing responsive and adaptive design. Like CSS, Flutter also provides media query objects with details regarding device size, orientation, pixel density, and so on. Like Android, we can decide to build different layouts based on screen size and orientation if we choose to. In this section, we will learn about those tools and techniques. There are basically two things we need to make designs responsive and adaptive: the viewport size and the platform or device that it's running on. Flutter provides ways to get this information. Apart from tools to provide this information, Flutter by default provides various widgets that allow you to lay out your design in a way that adapts to the changing viewport.

First, let's talk about the **MediaQuery** (`https://api.flutter.dev/flutter/widgets/MediaQuery-class.html`) object. Flutter provides the `MediaQuery` object that you can access in your application and also provides the **MediaQueryData** (`https://api.flutter.dev/flutter/widgets/MediaQueryData-class.html`) object, which has various information regarding current media – for example, a window that the application is running on. This is somewhat like the media query that is used in CSS if you have done web development before. As you can see, the `MediaQueryData` object provides a viewport size, view insets, view padding, platform brightness, and various other information regarding the device and viewport. You can use this information to make your application respond to the changes in the viewport.

Next, when we want to figure out the platform that the application is currently running on, we have a `Platform` as well as a constant called **kIsWeb**. The `Platform` class provides different properties to figure out which platform the app is running on. kIsWeb is a Boolean value that is `true` only when the application is running on the Flutter web platform. These handy properties allow us to figure out which platform and operating system our application is running on during the application's runtime. Once we know the platform details, we can make platform-specific decisions to modify and adapt our application. For example, in an application where the user has to choose between different options, you could use bottom sheets if the application is running on iOS, dialog if the application is running on Android, and a drop-down button if it's running on desktop and web.

Now, apart from these two tools, Flutter provides other widgets to make your application responsive. `AspectRatio`, `FittedBox`, `FractionallySizedBox`, `LayoutBuilder`, and so on are just a few of those widgets. We will use the aforementioned tools and some of these widgets in the next section to make our application responsive.

> **Important Note**
> While going through the chapter, remember that the method that is described in this chapter is not the only way to achieve responsive and adaptive design. There are always multiple ways to achieve the same thing.

There are some great open source example web applications that you can refer to see how to achieve responsive and adaptive design:

- Flokk (`https://github.com/gskinnerTeam/flokk`)
- Flutter gallery (`https://github.com/flutter/gallery`)

Making our app responsive and adaptive

In this section, we will use some of the tools and techniques we learned about in the previous section to improve our Flutter Academy app. We will make our Flutter Academy app responsive and adaptive.

Defining metrics for responsive design

First, we will start by defining some screen size constants. We will use some of the popular screen sizes on the market that are also used by other web application frameworks.

In `chapter3_start/lib/res`, create a file named `responsive.dart`. Inside this file, define the following class and the constants. As we are working on Flutter on the web, the sizes here are taken from Tailwind CSS, a popular CSS framework. You can use this or decide the sizes yourself by researching the common screen sizes:

```
class ScreenSizes {
  static const double xs = 480.0;
  static const double sm = 640.0;
  static const double md = 768.0;
  static const double lg = 1024.0;
  static const double xl = 1280.0;
  static const double xxl = 1536.0;
}
```

Now that we have the screen sizes, we will use this and the `MediaQueryData` we discussed in the previous section to first make our featured section widget responsive.

Updating a featured section to be responsive

Based on the screen size, we want to change the layout of a featured section widget. Before writing the code, we need to decide the layout. Currently, the featured section shows the image and the text in a row, each occupying 50% of the space using the expanded widget. So, let's say for a screen size above the medium (768), we want the current layout, whereas for a screen size below that, we want the information to be displayed in a single column. For this, we will need to remove the fixed width given to the top container and switch the top-level `Row` widget that we are using with the `Flex` widget.

The Flex widget is the primitive version of Row and Column and allows us to define the direction by laying out the children either horizontally or vertically. To decide the direction, we will get the screen width from the MediaQueryData object and compare it with the screen size constants that we have defined. We might want to use this strategy in a couple of widgets later on, so first, in chapter3_start/lib/res/responsive.dart, let's create a method called getAxis as follows:

```
Axis getAxis(double width) {
    return width > ScreenSizes.md ? Axis.horizontal :
      Axis.vertical;
}
```

This function takes a width and returns Axis. If the provided width is greater than the medium screen size that we have defined, then the axis will be horizontal; otherwise, it will be vertical. Let's update our FeaturedSection widget using the Flex widget along with the getAxis method we just defined. Update the build method as follows:

```
Widget build(BuildContext context) {
    final width = MediaQuery.of(context).size.width;
    return Container(
       height: width > ScreenSizes.md ? null : 600,
       padding: const EdgeInsets.all(32.0),
       child: Flex(
          direction: getAxis(width),
          children: [
             . . .
          ],
       ),
    );
}
```

The following figure shows how it looks on a large screen:

Figure 3.1 – The featured section on a large screen

The following figure shows how it looks on a small screen:

Figure 3.2 – The featured section on a small screen

Here, we are getting the width from the `MediaQuery`. Accessing the width from MediaQuery this way makes sure that whenever the width changes, the `build` method is called again with the new value for the width. Then, we are defining the container height instead of the width. We are giving a fixed height when the width is greater than the medium screen size we defined. This makes sure that when the widgets are laid out horizontally, we have a proper section with enough height. And finally, we replaced the `Row` widget with `Flex`, and the direction is defined again by the same `getAxis` method we defined previously. The `getAxis` method will decide whether to lay children down horizontally or vertically, based on the width given to it as a parameter. Now that our featured section looks great, on both small and large screens, let's move on to modifying the footer in the next section.

Modifying the footer

Next, we will work on the footer, as it involves similar decisions. So, let's get started. Like before, we first need to decide how it looks in different screen sizes. Right now, the footer has three different columns of links, and below them, there is a row with a title and copyright information. To make it simple, we will use the same logic as before – that is, when we are in screens a size larger than the medium size that we have defined, we will use the current layout; otherwise, we will lay out everything in a single column.

We will start by getting the width from MediaQuery inside our `build` method. Open `lib/widgets/footer.dart` and update the `build` method by adding the following lines:

```
Widget build(BuildContext context) {
  final width = MediaQuery.of(context).size.width;
  ...
}
```

Next, we will make a few modifications. Right now, we are giving the footer's wrapping container a height of `300`; however, when we switch to a single column layout on small screens, this will not be sufficient, and we will run into an overflow error. So, we will drop the height and also the expanded widget we have used for our row inside the top-level column in the footer widget.

The current `build` method code in the footer widget looks like this:

```
return Container(
  height: 300,
  color: Colors.grey.shade900,?
  child: Column(
    mainAxisAlignment: MainAxisAlignment.start,
    children: [
      const SizedBox(height: 20.0),
      Expanded(
```

```
          child: Row(
            crossAxisAlignment: CrossAxisAlignment.start,
            . . .
```

We will update it to the following:

```
return Container(
  color: Colors.grey.shade900,
  child: Column(
    mainAxisAlignment: MainAxisAlignment.start,
    mainAxisSize: MainAxisSize.min,
    children: [
      const SizedBox(height: 20.0),
      child: Row(
        crossAxisAlignment: CrossAxisAlignment.start,
        . . .
```

The **Column** widget usually takes all the available vertical space, and we run into an overflow error; however, by setting mainAxisSize as MainAxisSize.min, we tell the column to only take the minimum space required.

Now, we will replace the Row widget that we are using to display the three columns of links with the Flex widget as we did before. Here again, we will use the getAxis method we wrote previously to set the direction of our Flex widget, so that footer links are displayed in three columns on larger screens and a single column on smaller screens. Again, update the build method code as follows:

```
return Container(
  color: Colors.grey.shade900,
  child: Column(
    mainAxisAlignment: MainAxisAlignment.start,
    mainAxisSize: MainAxisSize.min,
    children: [
      const SizedBox(height: 20.0),
      child: Flex(
        direction: getAxis(width),
        crossAxisAlignment: CrossAxisAlignment.start,
        . . .
```

Similarly, inside the `Flex` children, we will display the sized box for spacing only when the screen width is larger than the medium and the `Flex` is laid out horizontally. So, we update the sized box as follows:

```
if (width > ScreenSizes.md) const SizedBox(width: 20.0),
```

This will make sure the sized box is only displayed when the screen width is greater than the medium screen size that we have defined. Similarly, we will do the same thing for the `Spacer` widget, as we don't want spacers when everything is laid out in a single column. Update the `Spacer` widget as follows:

```
if (width > ScreenSizes.md) const Spacer(),
```

In the same manner, let's also update the row below the links displaying the app name and the copyright information, as follows:

```
Flex(
  direction: getAxis(width),
  children: [
    Padding(
      padding: width > ScreenSizes.md
          ? const EdgeInsets.only(left: 30.0)
          : const EdgeInsets.all(0),
      child: Text(
        "Flutter Academy",
        style:
          Theme.of(context).textTheme.headline5?.copyWith(
              color: Colors.white,
            ),
      ),
    ),
    width > ScreenSizes.md
        ? const Spacer()
        : const SizedBox(height: 10),
    Padding(
      padding: width > ScreenSizes.md
          ? const EdgeInsets.only(right: 30.0)
          : const EdgeInsets.only(bottom: 10),
      child: Text(
        "© 2018 Flutter Academy",
        style: Theme.of(context)
```

```
            .textTheme
            .headline6
            ?.copyWith(color: Colors.white),
        ),
      ),
    ],
  ),
),
```

The following screenshot shows how the footer looks on a large screen:

Figure 3.3 – The footer on a large screen

The following screenshot shows how the footer looks on a small screen:

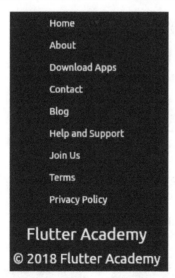

Figure 3.4 – The footer on a small screen

Here, we are again replacing the row with a `Flex` widget and using the `getAxis` method to determine the direction of it. We are also making sure that the `Spacer` widget is displayed on screens sized larger than medium. Finally, based on the screen size, we are also modifying the padding to make it look consistent.

We have now successfully applied the responsive design techniques we learned in the previous sections to make our featured section and footer responsive. Next, let's see how we can update a design based on the input that users provide.

Making our app input-ready

Next, let's talk about input. The web runs on desktop as well as mobile devices. Unlike desktops, mobile devices have touch input. Therefore, it is better to use large areas for intractable items such as buttons. So, let's handle that in our call-to-action section. Our call-to-action button has the same size on all devices. So again, using the media query, we will decide on our button size. On devices smaller than the medium screen size we defined, which we assume are tablets and mobiles with touch input, we will make the button large so that it's easy to tap, and on large screens, we will make it smaller.

Open `lib/widgets/call_to_action.dart` and update the Elevated Button widget, like so:

```
ElevatedButton(
  style: ElevatedButton.styleFrom(
    fixedSize: MediaQuery.of(context).size.width >
      ScreenSizes.md
        ? Size(180, 50)
        : Size(180, 60),
  ),
  onPressed: () {
    print("register");
  },
  child: Text("Get Started"),
)
```

Here, we just updated the size of the button based on the screen size. You can modify the buttons on the featured section using the same principle.

This is how the button looks on small and large screens:

Figure 3.5 – The button on small and large screens

Now that we have seen how we can modify the button based on the input that might be available, making it easier for the user to respond and provide the required input, let's also update the app navigation so that it's easily accessible in each screen size.

Updating navigation to be responsive

Now, it's time to handle the top navigation. Unlike other layouts we have dealt with, we will use a totally different idea here. For the top navigation, on a large screen, we will display the horizontal navigation as it is right now. It is displayed using the actions in `AppBar`:

Figure 3.6 – Top navigation on a large screen

On a small screen, we want to hide the top navigation so, let's modify the `AppBar` actions like so:

```
AppBar(
  ...
  actions: (MediaQuery.of(context).size.width <=
    ScreenSizes.md)
      ? null
      : [
          TextButton(
            child: Text("Home"),
            style: TextButton.styleFrom(
              primary: Colors.white,
            ),
            onPressed: () {},
          ),
          TextButton(
            child: Text("Courses"),
            style: TextButton.styleFrom(
              primary: Colors.white,
            ),
            onPressed: () {},
          ),
          TextButton(
            child: Text("About"),
            style: TextButton.styleFrom(
              primary: Colors.white,
            ),
            onPressed: () {},
          ),
```

```
        TextButton(
          child: Text("Contact"),
          style: TextButton.styleFrom(
            primary: Colors.white,
          ),
          onPressed: () {},
        ),
      ],
    ),
  ),
```

So, we just hid the actions on a screen size smaller than medium.

However, on the small screen, we want to display side navigation, which is hidden by default and only displayed when the user taps the icon on the top bar. In order to display this menu, Material Design provides a `drawer` property on the scaffold. We will use that to build a menu that looks like this:

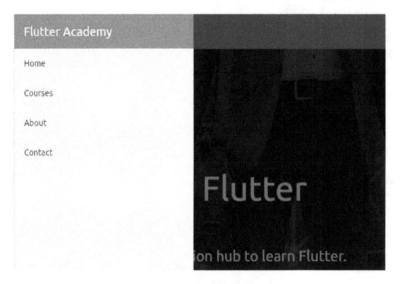

Figure 3.7 – Side navigation on a small screen

In the scaffold, let's add `drawer`, as shown in the following code block:

```
return Scaffold(
  body: ListView(
    ...
  )
  drawer: MediaQuery.of(context).size.width >
```

```
ScreenSizes.md
  ? null
  : Drawer(
      child: ListView(
        children: [
          Container(
            color: Theme.of(context).primaryColor,
            padding: const EdgeInsets.all(16.0),
            child: Text(
              "Flutter Academy",
              style: Theme.of(context)
                    .textTheme
                    .headline6
                    ?.copyWith(color: Colors.white),
            ),
          ),
          ListTile(
            title: Text("Home"),
            onTap: () {},
          ),
          ListTile(
            title: Text("Courses"),
            onTap: () {},
          ),
          ListTile(
            title: Text("About"),
            onTap: () {},
          ),
          ListTile(
            title: Text("Contact"),
            onTap: () {},
          ),
        ],
      ),
    ),
);
```

Here, again, we decided that we want to display `drawer` only when the screen size is smaller than the medium size. Now, we have our responsive navigation too.

This is all for now. We will learn more about these techniques and use more of these widgets further as we proceed to implement more features in our application.

Summary

In this chapter, we learned about responsive and adaptive design. We learned why we need to make our applications responsive and adaptive. We learned that as Flutter is a cross-platform framework, we need responsive and adaptive designs more than ever. Finally, we used the information that we gathered during the lesson to make our Flutter Academy app's home page responsive and adaptive.

In the next chapter, we will learn how Flutter on the web works under the hood and how to use that information to benefit our application.

Part 2: Flutter Web under the Hood

In this part, you will learn how to build scalable web apps with the native look and feel of the web platform using Flutter.

This part comprises the following chapters:

- *Chapter 4, Flutter Web under the Hood*
- *Chapter 5, Understanding Routes and Navigation*
- *Chapter 6, Architecting and Organizing*

4

Flutter Web under the Hood

In the previous chapter, we learned why we should make our applications responsive and adaptive. We also updated our home page layout to make it responsive and adaptive. In this chapter, we will go deep inside the Flutter Engine and learn how it works under the hood. We will learn how Flutter produces apps for different platforms from the same code base, especially for the web.

By the end of this chapter, you will have learned about the different types of renderers available for building web applications from Flutter code, and the advantages and shortcomings of those renderers. You can use this knowledge to build your application using different renderers based on your requirements.

In this chapter, we will cover the following main topics:

- How does the Flutter Engine produce web apps?
- Different types of web renderers and their advantages and disadvantages.
- Choosing between HTML and CanvasKit renderers.

Technical requirements

The technical requirements for this chapter are as follows:

- Flutter version 3.0 or later installed and running
- Visual Studio Code or Android Studio
- Google Chrome browser

You can download the code samples for this chapter from the book's official GitHub repository at `https://github.com/PacktPublishing/Taking-Flutter-to-the-Web`. The starter code for this chapter can be found inside the `chapter4_start` folder.

How does the Flutter Engine produce web apps?

In the previous chapters, we discussed the fact that Flutter enables us to build applications for multiple platforms from the same code base. So, the same Flutter code written in Dart can be compiled to any supported platform.

Flutter is essentially a toolkit for building user interfaces, but it also allows applications to interface directly with underlying platform services. When we build a Flutter application for release, the app code is compiled directly to machine code. For example, for the web, a Flutter app is compiled into JavaScript. Let's learn more about the Flutter architecture in order to understand how a Flutter app is compiled into web, or any other platform-specific machine, code. The following figure shows an overview of the Flutter framework's architecture:

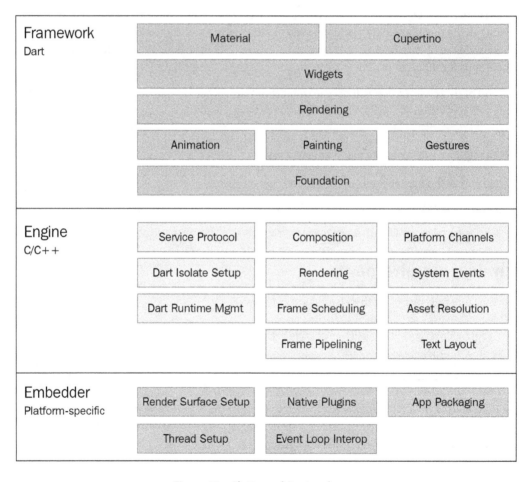

Figure 4.1 – Flutter architecture layers

Flutter is designed as an extensible, layered system. As you can see in *Figure 4.1*, there are three distinct layers. Firstly, we have the framework, which is written in Dart, then the engine, written in C/C++, and finally, the embedder, which is platform-specific. Each layer of Flutter exists as an independent library that depends on the underlying layer. At the framework level, every part is designed as optional and replaceable. For all the other platforms, the layered architecture displayed in *Figure 4.1* is used. However, the web platform is essentially different from other platforms.

For web support, Flutter had to provide a reimplementation for the Flutter Engine on top of standard browser APIs. Currently, there are two options in Flutter for rendering Flutter content on the web. They are HTML and WebGL. To render content in HTML mode, HTML, CSS, Canvas, and SVG are used by Flutter, whereas in WebGL mode, a version of Skia compiled to WebAssembly called CanvasKit is used. We will look at both of these renderers in detail in the next section.

The web version of the architectural layer diagram is displayed in *Figure 4.2*. The most notable difference is that we don't require a Dart runtime. Instead, the Flutter framework (along with user code) is compiled to JavaScript. Most developers would never create a line of code that has semantic differences across platforms because Dart has relatively few semantic variations across all of its modes (JIT versus AOT, native versus web compilation):

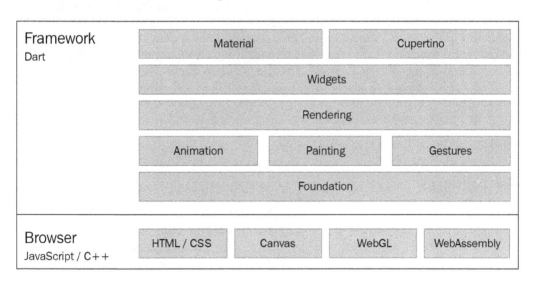

Figure 4.2 – Flutter web layered architecture

Flutter on the web employs the `dartdevc` compiler during development, which offers incremental compilation and supports hot restart (Hot reload is not currently supported in Flutter <= 3.0). However, Flutter is always evolving, and many improvements are made with each release. So, by the time you are reading this, there might be a lot more updates to the Flutter web version. To learn what's going on in Flutter on the web, you can visit `https://flutter.dev/multi-platform/web`. Flutter uses dart2js, Dart's highly efficient production JavaScript compiler, when you're ready to build your Flutter web application for production. Flutter's core and framework layer (along with your application) are packaged by dart2js into a single minified source file that can be delivered to any web server. If you want, you can set to split the code into multiple files through deferred imports. So, this is how the Flutter Engine compiles Dart code and produces a web application from our Flutter code. Next, we will learn about the different types of web renderers available in Flutter and their advantages and disadvantages.

Different types of web renderers and their advantages and disadvantages

When running and building apps for the web using Flutter, we can choose between two different renderers. In this section, we will learn what those renderers are and what the advantages and disadvantages of using each of them are.

So, what are renderers? As the name implies, they are used to render our Flutter app on the web. The two renderers that we can use to render our Flutter application on the web are the HTML renderer and the CanvasKit renderer. So, how did these renderers come to be? As we discussed in the previous section, to enable web support, the Flutter team had to rewrite the Flutter Engine on top of browser APIs. The first solution was to use HTML, CSS, and CanvasAPI on the web, which is how the HTML renderer came to be. This was the easier (and first) solution. However, later came another solution: using WebGL. For this, the team brought Skia (the underlying graphic engine that supported Flutter on other platforms to the web) via CanvasKit (`https://skia.org/docs/user/modules/canvaskit`) and WebAssembly. Unlike the HTML renderer, CanvasKit doesn't depend on HTML or CSS, making it independent of browser-rendering techniques. This also allowed the Flutter application to have more control over the screen pixels, enabling pixel-perfect rendering on the web.

Let's look at each of the renderers in detail.

HTML renderer

Flutter's HTML renderer uses HTML elements, CSS, Canvas elements, and SVG elements, that is, a combination of these four elements, to render a user interface built using Flutter. This was the first solution devised by the Flutter team to port Flutter to the web. This renderer has a smaller download size and is suitable when prioritizing download size over performance.

The advantages of using the HTML renderer are as follows:

- It has a smaller bundle size than CanvasKit and therefore, the initial load is faster.
- It uses native text rendering, allowing for the use of system fonts in a Flutter application.

Some of its disadvantages are as follows:

- Fidelity issues with text layout

- Less performant than CanvasKit

- Problematic SVG support (SVG is supported on the web natively; however, on other platforms, it requires the `flutter_svg` package, so you need to use the **Image** widget on the web and the **SvgPicture** widget on other platforms)

Text layout was one of the biggest issues the Flutter team faced with web support, and it still is an issue. That is why in *Chapter 1, Getting Started with Flutter on the Web*, we suggested that Flutter is not yet suitable for text-heavy websites.

CanvasKit renderer

As the CanvasKit renderer uses the web port of Skia, this renderer is fully consistent across Flutter mobile and desktop and has the best performance. CanvasKit provides the fastest path to your browser's graphics stack and offers higher graphical fidelity, but adds about 2 MB in download size.

The advantages are as follows:

- Blazing fast and extremely performant

- Accurate text measurement and layout

- Behaves pretty much the same as Flutter for mobile/desktop (all paint methods are supported and SVGs work as normal)

The disadvantages are as follows:

- Does not use native text rendering. Therefore, custom fonts must be shipped along with it.

- Emoji preloading issues. A custom font for emojis must be preloaded into your app when it is first loaded. The font (Noto Color Emoji) is also quite large (9 MB at the time of writing).

- Larger bundle size. The renderer is about 2 MB larger than the HTML renderer.

- **Cross-Origin Resource Sharing (CORS)** issues.

Both renderers have great advantages. However, their disadvantages are just as great as their advantages. To choose the best renderer for your application, you will have to consider a few things. We will investigate these in the next section.

Choosing between the HTML and CanvasKit renderers

We should consider our requirements in order to decide between the HTML and CanvasKit renderers. In this section, we will explore our requirements and accordingly choose the renderer. Additionally, we will also look into different options available to set up these renderers.

Loading time

If your application requires faster initial loading times, especially on mobile, you should go with the HTML renderer. As we discussed in the previous section, the CanvasKit renderer adds an additional 2 MB download size. The default behavior is to select the HTML renderer on mobile, whereas the CanvasKit renderer would be selected on desktops.

Data usage

Firstly, CanvasKit's bundle size is larger than HTML's. Also, if you're going to use emojis in your app, you won't want the user to download a large font in addition to your application. So again, if you're concerned about data consumption, you should go with the HTML renderer.

Text fidelity

If your app has a lot of text (such as a note-taking or journal app), CanvasKit would be a better option. However, if your app is dealing with a lot of emojis, it might be safer to stick with the HTML renderer for now.

Performance and uniformity

If you are concerned about high performance, and your application is graphic-centric, or if it relies on canvas APIs available only on CanvasKit, the CanvasKit renderer would be the default choice. For example, if you are building a graphic design tool, then CanvasKit would provide the best output.

Most of the time, however, we may not need the performance CanvasKit offers. On these occasions, it may be worth trading performance for faster loading times and a smaller bundle size.

Now that we know how to decide which renderer to use when, let's look at what options are available to set the default renderer.

Command-line options

The Flutter CLI has the `--web-renderer` command-line option, which takes any one of the following values:

- `auto`
- `html`
- `canvaskit`

`auto` is the default option. This option automatically chooses which renderer to use. When the app is running in a mobile browser, it will choose the HTML renderer, whereas the CanvasKit renderer is chosen when the app is running in a desktop browser. `html` will always use the HTML renderer, and `canvaskit` will always use the CanvasKit renderer.

This flag can be used with the `run` or `build` subcommands. Let's test it with our application. Navigate to the `flutter_academy` project folder in your terminal, and let's run the application with the following code:

```
flutter run -d chrome –web-renderer html
```

Similarly, you can set the web renderer as CanvasKit while running. You might not notice any difference in an application such as this.

Let's also see how we can build with different renderers. First, let's build with the CanvasKit renderer. Use the following command:

```
flutter build web --web-renderer canvaskit
```

You can run and test the app built with the CanvasKit renderer. Similarly, you can change to the HTML renderer and test the application. As we move further through the chapter and add more features to our application, the difference in performance might become more apparent, so we will test our application with different renderers in the upcoming chapters as well.

Runtime configuration

There is also a runtime configuration, using which you can switch between different renderers at runtime and allow your users to choose the renderer for your application. To override the web renderer at runtime, follow this process:

1. First, build the application with the `auto` option for `–web-renderer`.
2. Then, insert a `<script>` tag in the `web/index.html` file before the `main.dart.js` script, where you can set `window.flutterWebRenderer` to "canvaskit" or "html":

```html
<script type="text/javascript">
  let useHtml = // some code to decide when to use
                // HTML and CanvasKit renderer
  if(useHtml) {
    window.flutterWebRenderer = "html";
  } else {
    window.flutterWebRenderer = "canvaskit";
  }
</script>
<script src="main.dart.js"
  type="application/javascript"></script>
```

This must happen before loading `main.dart.js`. Once the Flutter Engine startup process begins in `main.dart.js`, the web renderer can't be changed.

Summary

In this chapter, we looked at the architecture of the Flutter framework and learned how Flutter produces multiplatform applications. We looked at how Flutter produces web applications. We also learned about the different renderers available to build and run web apps with Flutter and learned about the advantages and disadvantages of both of those renderer options. Finally, we looked at how we could choose between the two renderers and when to choose which renderer. I hope this information will help you choose the most suitable renderer based on your target user and application requirements. In the next chapter, we will learn about routing and navigation in Flutter on the web.

Understanding Routes and Navigation

In the previous chapter, we learned how Flutter works under the hood to produce applications for different platforms, especially for the web. We also learned about different renderers available to build web apps, and how and when to use them. In this chapter, we will start learning about routing and navigation. We will also implement navigation in our Flutter Academy application.

By the end of this chapter, you will understand how to use routes and how to navigate between different screens. You will also learn how to use these principles to add navigation to our application. You will learn about the Navigator 2.0 API and how it's useful for web applications. We will then change our application to use Navigator 2.0, allowing us to use navigation effectively, which will include parsing URLs received from the web browser and updating the navigation stack based on the URL.

In this chapter, we will cover the following main topics:

- The basics of navigation in Flutter
- An introduction to Navigator 2.0
- Implementing Navigator 2.0 in our app

Technical requirements

The technical requirements for this chapter are as follows:

- Flutter version 3.0 or later installed and running
- Visual Studio Code or Android Studio
- Google Chrome browser

You can download the code samples for this chapter and other chapters from the book's official GitHub repository at `https://github.com/PacktPublishing/Taking-Flutter-to-the-Web`. The starter code for this chapter can be found inside the `chapter5_start` folder.

The basics of navigation in Flutter

Any production-level application requires multiple pages on which to display information. In order to manage multiple pages, we need to provide proper navigation support for users. Flutter provides different APIs for navigation. In this section, we will look at the imperative navigation API provided by an early version of Flutter, which is still effectively used. There are some shortcomings of this type of navigation, which is why the new declarative navigation API was introduced. We will look into these shortcomings, and in the next section, we will look at the new declarative navigation API.

To use Flutter's imperative navigation, we need to understand two terms:

- **Navigator**: `Navigator` is a widget that manages a stack of `Route` objects.
- **Route**: Route is an object managed by `Navigator` that represents a screen. This is typically implemented by classes such as `MaterialPageRoute`.

When using imperative navigation, we use various static methods provided by `Navigator` to push and pop routes in Navigator's stack. The routes can be anonymous or named. Anonymous routes, which are still the most suitable for mobile applications, can display screens on top of each other like a stack, and can easily be achieved using Navigator.

`MaterialApp` and `CupertinoApp` already use Navigator under the hood. Navigator can be accessed using `Navigator.of()` or a new screen can be displayed using `Navigator.push()`, and by using `Navigator.pop()`, we can return to the previous screen.

As we are using `MaterialApp`, we can easily use this Navigator API. Let's see how to navigate to the **About** page using simple anonymous route:

1. Open `lib/widgets/top_nav.dart` and in `TextButton` with the **About** text, update the `onPressed` action as follows:

   ```
   onPressed() {
       Navigator.of(context).push(
         MaterialPageRoute(builder: (context) =>
           AboutPage()));
   }
   ```

2. Make sure to import `lib/pages/about.dart`.

   If you run the project and tap on the **About** link on the top navigation, you should see the following page:

Figure 5.1 – About page

However, you will notice that the URL in the address bar has not changed. When push() is called, AboutPage is placed at the top of the HomePage widget. So, this route, which we thought very suitable for mobile applications, is not suitable for the web for displaying new pages. This is because on the web, we tend to associate pages with specific URLs. Next, we will implement this same route with named routes. Let's see how:

1. First, open lib/main.dart and update MaterialApp, as follows:

```
return MaterialApp(
  title: 'Flutter Demo',
  debugShowCheckedModeBanner: false,
  theme: ThemeData(
    primarySwatch: Colors.blue,
  ),
  routes: {
    '/': (_) => HomePage(),
    '/about': (_) => AboutPage(),
  },
);
```

Here, we are adding the route's properties with a list of named routes and associated pages.

2. Now, let us update the navigation code we wrote in lib/widgets/top_nav.dart, as follows:

```
OnPressed() {
  Navigator.of(context).pushNamed("about");
}
```

This will also navigate to about as in the previous code, but if you look at the URL now, you can see that it has changed.

Figure 5.2 – Named URL route

In a similar way, we can add named routes for other pages of our application.

3. Let's update the route properties of `MaterialApp` as follows in `lib/main.dart`:

```
routes: {
  '/': (_) => HomePage(),
  '/about': (_) => AboutPage(),
  '/contact': (_) => ContactPage(),
  '/courses': (_) => CoursesPage(),
},
```

We have added routes for new pages. We can also update the top navigation to add navigation to each link.

4. Update the top navigation actions, as follows:

```
actions: (MediaQuery.of(context).size.width <=
ScreenSizes.md)
  ? null
  : [
      TextButton(
        child: Text("Home"),
        //...
        onPressed: ()  => Navigator.pushNamed(context,
                                               '/');
      ),
      TextButton(
        child: Text("Courses"),
        //...
        onPressed: () => Navigator.pushNamed(context,
          '/courses');
      ),
      TextButton(
        child: Text("About"),
        //...
        onPressed: () => Navigator.pushNamed(context,
          '/about');
      ),
      TextButton(
        child: Text("Contact"),
```

```
//...
onPressed: () => Navigator.pushNamed(context,
  '/contact');
),
],
```

5. Similarly, update the onTap action for each ListTile widget on the drawer navigation, adding the relevant navigation as we did with our top navigation.

Now, if you run the application, you can navigate to each page and the URL also gets updated. Also, if you directly update the URL on the address bar to a specific page, that page will be displayed as the initial page. This is great. We can use static values in the URLs to navigate to a specific page. Now, what happens when you want to add dynamic parameters to URLs? Right now, you can get to the **Courses** page and tap on an individual course, and you can navigate to the **Course details** page. However, say you want to navigate to a different course by passing the ID in the URL, like so: details/:id. The current named route doesn't allow us to parse those parameters. Next, we will look into how we can implement an advanced named route using onGenerateRoute, which will allow us to parse the URL parameters.

Let's remove the routes parameter from the MaterialApp widget and instead use the onGenerateRoute parameter. Update the MaterialApp widget in lib/main.dart, as follows:

```
return MaterialApp(
  title: 'Flutter Demo',
  debugShowCheckedModeBanner: false,
  theme: ThemeData(
    primarySwatch: Colors.blue,
  ),
  onGenerateRoute: (settings) {
    return MaterialPageRoute(
      builder: (_) {
        switch (settings.name) {
          case '/':
            return HomePage();
          case '/about':
            return AboutPage();
          case '/contact':
            return ContactPage();
          case '/courses':
            return CoursesPage();
```

```
        default:
          final pathSegments =
            Uri.parse(settings.name!).pathSegments;
          print(pathSegments);
          if (pathSegments.length == 2 &&
              pathSegments[0] == 'courses') {
            final courseId =
              int.parse(pathSegments[1]);
            return CourseDetailsPage(courseId:
            courseId);
          }
          return Error404Page();
        }
      },
    settings: settings,
  );
  },
);
```

As you can see, we get name as the RouteSettings name parameter, and we can manipulate and test name as we like. So, using the preceding code, we are handling simple named routes as well as advanced routes by parsing the route name.

We are parsing the name parameter using Uri.parse. Then, we are getting the path segments from the parsed URI. By testing different parameters received in the path segments, we set the page for the route. Now, if you run the application and directly type the URL with a course ID in the URL segment in your address bar, you will be navigated to that specific course's details page.

However, this imperative navigation style has some limitations. We are not able to update the navigation stack as we please. Updating the navigation stack dynamically based on application state (which is required for applications on the web) is not possible with this method. Also, there is no proper way to handle the navigation received from a native platform and build a proper navigation stack based on that. To learn more about this style of navigation in Flutter, you can find more resources in the official documentation at https://docs.flutter.dev/cookbook/navigation. Navigator 2.0, which is a declarative navigation API, was introduced to tackle these issues. We will learn more about Navigator 2.0 in the next section.

An introduction to Navigator 2.0

Navigator 2.0 is a declarative navigation API that allows us complete control over our application's navigation stack. This also allows us to control routes much like any other state of our application.

The Navigator 2.0 API adds new classes to the framework in order to make the app's screens a function of the app state. This also provides the ability to parse routes from the underlying platform (such as web URLs). Navigator 2.0 is provided alongside the existing navigator. That means Navigator 2.0 is not a replacement for the existing imperative navigation API; it's just another way of handling our app's navigation. It was introduced because with Flutter being a cross-platform framework, the imperative navigation API (which is more suitable for mobile-only applications) was limiting the progress of Flutter on the web. Navigator 2.0 is more suitable for platforms such as applications on the web, where underlying platform routes are more frequent and should be handled properly. Alongside the existing navigation API, the following resources were added to the framework to support the new navigation API:

- **Page**: `Page` is an immutable object used to set the navigator's stack. The navigator's stack is just a list of `Page` objects.

- **Router**: `Router` configures the list of pages in the navigator's stack. This list of pages typically varies when the app's state or the underlying platform changes, and it is this list that controls what is presented on the screen.

- `RouteInformationParser`: In order to manage the navigation state, the route information parser accepts `RouteInformation` from `RouteInformationProvider` and parses it into a user-defined data type.

- `RouterDelegate`: The app-specific behavior of how `Router` obtains information about the app's state and how it manages it is defined by the router delegate. It listens to `RouteInformationParser` and the app state before constructing Navigator with the current stack of pages.

- `BackButtonDispatcher`: The back button dispatcher reports back button presses to Router.

The following diagram shows how `RouterDelegate` interacts with Router, `RouteInformationParser`, and the app's state:

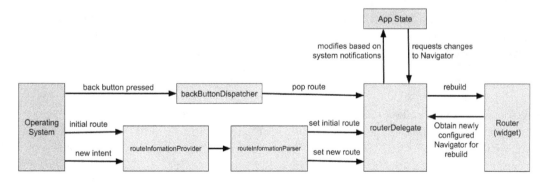

Figure 5.3 – Navigator 2.0 (source: https://docs.google.com/document/
d/1Q0jx0l4-xymph9O6zLaOY4d_f7YFpNWX_eGbzYxr9wY)

Here's an example of how these components work together.

`RouteInformationParser` translates a new route (for example, `courses/flutter-beginners`) that the platform emits into an abstract data type that you create in your app (for example, `Uri`).

When this data type is sent to the `setNewRoutePath` method of `RouterDelegate`, it must update the application state to reflect the change (for example, by adding a course details page to the pages list and calling `notifyListeners`).

`RouterDelegate` rebuilds using its `build()` method when `notifyListeners` is called.

The `Navigator` widget with updated pages is returned by `RouterDelegate.build()`. The pages returned reflect the change to the app state (for example, a new page that shows course details is added to the pages list).

This is how Navigator 2.0 works under the hood. You can also learn more about Navigator 2.0 at `https://medium.com/flutter/learning-flutters-new-navigation-and-routing-system-7c9068155ade`. In order to implement Navigator 2.0 in our application, we will have to implement these abstract classes in our application and handle the navigation state. We will do that in our Flutter Academy application in the next section.

Implementing Navigator 2.0 in our app

Let's start. As we are using Material Design and `MaterialApp`, for simplicity, we will not be implementing our own `Page` object, we will instead use `MaterialPage` provided by Flutter's material library. If you wish, you can implement `Page` yourself by creating a new class that extends `Page` and implementing the required functionality. So, we will start by implementing `RouterDelegate`. `RouterDelegate` requires a generic type that defines our route. Here, we will be using a **Uniform Resource Identifier** (**URI**) as it is powerful as well as simple. With a URI, we can provide advanced paths, as well as manipulate them easily. This is even more suitable for the web, as in the browser, we are always dealing with URIs. Inside `lib/`, create a folder named `routes`, and inside that folder, create a file called `router_delegate.dart`. First, we will create our class by extending `RouterDelegate`, setting URI as the generic type, like so:

```
class AppRouterDelegate extends RouterDelegate<Uri> {}
```

If we look at the source code of `RouterDelegate` by pressing *Ctrl/Cmd* and clicking on the class, we can see that it's an abstract class and has a bunch of functions that we must implement. Among them, we must implement `build` and `setNewRoutePath`. There are three other methods that we must implement, `popRoute`, `addListener`, and `removeListener`, for which we will be using existing mixins already provided by Flutter. Let's go through these methods and implement them one by one.

First, let's start by implementing the `build` method. The `build` method of `RouterDelegate` should return a `Navigator` widget, which provides a list of pages defining the application's route stack:

```
@override
Widget build(BuildContext context) {
  final pages = _getRoutes(_path);
  return Navigator(
    pages: pages,
  );
}
```

In order to get the list of pages, we are calling `_getRoutes`, which we will implement in a bit. Before that, we also need to implement `onPopPage`. This method is called when pop is invoked and receives `route` and `result`. Route corresponds to a page found in the pages list we have. `result` argument is a boolean value with which the route is to complete. We should call `Route.didPop` and return whether this pop is successful from this method. The `Navigator` widget should rebuild, and as a result, the new `pages` list should not contain the page for the given route. Update the `Navigator` widget by adding the `onPopPage` method, as follows:

```
onPopPage: (route, result) {
  if (!route.didPop(result)) {
    return false;
  }
  if (pages.isNotEmpty) {
    _path = _path.replace(
        pathSegments: _path.pathSegments
            .getRange(0, _path.pathSegments.length - 1));
    notifyListeners();
    return true;
  }
  return false;
},
```

The next method we need to implement is `popRoute`, and for this, we will be using `PopNavigatorRouterDelegateMixin`, so update the class definition for our router delegate, as follows:

```
class AppRouterDelegate extends RouterDelegate<Uri>
    with PopNavigatorRouterDelegateMixin<Uri> {
```

This mixin provides an implementation for popRoute so we don't have to handle it ourselves. Next, in order to implement addListeners and removeListeners, we will use the ChangeNotifier mixin, so update the class definition, as follows:

```
class AppRouterDelegate extends RouterDelegate<Uri>
    with ChangeNotifier, PopNavigatorRouterDelegateMixin<Uri> {
```

The only method we need to implement next is setNewRoutePath. But before doing that, we will implement a go method, which will allow us to navigate to different pages from our application by passing a path as a string. Let's do that as shown in the following code:

```
go(String path) {
   this._path = Uri.parse(path);
   notifyListeners();
}
```

With this, we should also add Uri _path as a class property, as follows:

```
Uri _path = Uri.parse('/');
```

Let us now implement the setNewRoutePath method. This method is called by the router whenever our router information provider reports that a new route has been pushed to the application by the operating system. For web applications, whenever a new URL is entered in the address bar, the following method is called:

```
@override
Future<void> setNewRoutePath(Uri configuration) async =>
    go(configuration.toString());
```

Here, we are calling the go method we implemented previously in order to update the path property so that our navigation stack is then updated according to the path. Finally, we will implement the most important method that parses the path and then provides the list of pages for the Navigator widget. We will implement the _getRoutes method that we are calling in the build method. Add the _getRoutes method, as shown in the following code:

```
List<Page> _getRoutes(Uri path) {
   final pages = [
     MaterialPage(child: HomePage(), key: ValueKey('home')),
   ];
   if (path.pathSegments.length == 0) {
     return pages;
   }
}
```

```
switch (path.pathSegments[0]) {
  case 'contact':
    pages.add(MaterialPage(
      key: ValueKey('/contacts'),
      child: ContactPage(),
    ));
    break;
  case 'about':
    pages.add(MaterialPage(
      key: ValueKey('/about'),
      child: AboutPage(),
    ));
    break;
  case 'courses':
    pages.add(MaterialPage(
      key: ValueKey('/courses'),
      child: CoursesPage(),
    ));
    break;
  default:
    pages.add(MaterialPage(child: Error404Page(),
      key: ValueKey('error')));
    break;
}
if (path.pathSegments.length == 2) {
  if (path.pathSegments[0] == 'courses') {
    pages.add(
      MaterialPage(
        key: ValueKey('course.${path.pathSegments[1]}'),
        child: CourseDetailsPage(
          courseId: path.pathSegments[1],
        ),
      ),
    );
  } else {
    pages.add(MaterialPage(child: Error404Page(),
```

```
        key: ValueKey('error')));
    }
  }
  return pages;
}
```

This is the method that we are using in our preceding `build` method, that is, parsing the path and returning the appropriate list of pages required by the `Navigator` widget. This is the method where we manipulate pages and routes based on the navigation path.

Now, one final thing we need to implement is the `currentConfiguration` getter. This is called when our app's route information is changed. This method is responsible for Flutter web apps updating the URL in the navigation bar when the route changes. Our router delegate is now complete. You can find the complete code at `https://github.com/PacktPublishing/Taking-Flutter-to-the-Web/blob/main/Chapter05/chapter5_final/lib/routes/router_delegate.dart`.

Next, we will implement the parser for the routes. The route parser has to implement `RouteInformationParser`, which is a delegate that is used by `Router` to parse the route information into the configuration type defined by us. In our case, we are simply using a URI, so this should parse the available route information into a URI. This is the class used when the router is first built and any subsequent times when new route information is provided by `routeInformationProvider`. Create a new file named `app_route_parser.dart` inside `lib/routes/` and update it, as follows:

```
import 'package:flutter/material.dart';
class AppRouteInformationParser extends
RouteInformationParser<Uri> {
  @override
  Future<Uri> parseRouteInformation(RouteInformation
    routeInformation) async =>
      Uri.parse(routeInformation.location!);

  @override
  RouteInformation restoreRouteInformation(
    Uri configuration) =>
      RouteInformation(location: configuration.toString());
}
```

Now that we have our router delegate and route information parser implemented, we should update `MaterialApp` to use the new navigator instead. We should keep in mind that we can still use old navigation together with this; however, in our case, we will replace the old navigation with the new one. Let's get started:

1. First, open `lib/main.dart` and create instances of our router delegate and route information parser before the MyApp widget starts, as shown in the following code:

```
final routerDelegate = AppRouterDelegate();
final _routeParser = AppRouteInformationParser();
```

Here, we are making the instance of the router delegate public and the route information parser private, as we will require the router delegate to perform navigations from our code.

2. Now, update `MaterialApp`, as follows:

```
return MaterialApp.router(
  title: 'Flutter Demo',
  debugShowCheckedModeBanner: false,
  theme: ThemeData(
    primarySwatch: Colors.blue,
  ),
  routerDelegate: routerDelegate,
  routeInformationParser: _routeParser,
)
```

Here, we have made three changes:

- For `MaterialApp`, we are now using the `MaterialApp.router` constructor.

- We are passing the `routerDelegate` and `routeInformationParser` parameters as the instances we created earlier.

- We removed the `routes` and `onGenerateRoute` parameters as we are now handling our navigations through the new navigator.

Now that our `MaterialApp` knows how to handle routes using the delegates we just provided, we can use the `routerDelegate` instance we created in this chapter to navigate, instead of the existing imperative navigation. So, wherever we are calling the imperative navigator's pushNamed function, we can replace it with `routerDelegate.go()`. We pass the same path to the `routerDelegate.go()` method as we were passing to the pushNamed method.

For example, open `lib/widgets/top_nav.dart` and in the courses button's onPressed function, replace `Navigator.pushNamed(context, '/courses')` with `routerDelegate.go('/courses')`. Similarly, replace all the navigation with `routerDelegate.go()`. Remember to import `routerDelegate` from `main.dart`. You can similarly replace the navigation code in `lib/widgets/drawer.dart` as well.

With this change, we should now be able to build and run the application as we did previously, while using named routes. We should be able to access the URL directly or navigate by clicking on the navigation provided on the page. If you have any issues, compare your code against the code in `https://github.com/PacktPublishing/Taking-Flutter-to-the-Web/tree/main/Chapter05/chapter5_final`.

Summary

In this chapter, we looked at the different ways of implementing navigation in our application. We also learned about the benefits and shortcomings of each method. We also learned how to implement both types of navigation (imperative and declarative) in our application.

In the next chapter, we will learn about architecting and organizing our Flutter application. We will learn about the need for proper architecture and implement a scalable architecture in our Flutter application.

6

Architecting and Organizing

In the previous chapter, we learned about different ways of implementing navigation in our Flutter application. In this chapter, we will learn about the need for proper architecture and then organize our Flutter application into a simple yet scalable architecture. We will also learn how to organize our source files and folders to make the application easier to scale and work with.

By the end of this chapter, you will understand the need for proper architecture in an application. You will also learn about the **Model-View-View Model** (**MVVM**) architecture by implementing it within your Flutter application. You will also understand the need for proper source file organization and organize the files in your application for scalability and easy working.

In this chapter, we will cover the following topics:

- The need for properly scalable architecture
- Organizing files and folders
- MVVM architecture
- Implementing MVVM architecture

Technical requirements

The technical requirements for this chapter are as follows:

- Flutter version 3.0 or later installed and running
- Visual Studio Code or Android Studio
- Google Chrome browser

You can download the code samples for this chapter from the book's official GitHub repository at `https://github.com/PacktPublishing/Taking-Flutter-to-the-Web`. The starter code for this chapter can be found inside the `chapter6_start` folder.

The need for properly scalable architecture

We are in a world where business concepts and processes are constantly changing. As mobile application developers, it's up to us to develop ways to evolve our application architecture to match the ever-changing business concepts and processes. If our architecture is not able to evolve, the application will not be structurally sound and we will have to deal with a dreadful architecture made up of different components that are difficult to update and maintain.

An application's architecture is made up of all of the software modules and components, as well as the internal and external systems and their interactions. A well-designed application architecture guarantees that your apps can expand as per the business needs, satisfying the intended business and user objectives while ensuring that all ideas are properly segregated and have solid relationships. Scalable code ensures that you have the necessary architecture in place to add new features to an application without causing it to break.

As you update and add new requirements to your software project, and supposing you disregard the architectural side of things, you'll wind up with a spaghetti architecture, a maze of unmanageable synchronization and dependencies between various portions of your project.

The difficulty with spaghetti architecture is that it causes a slew of problems, the most serious of which are the following:

- **Poor service abstraction**: When services are not properly isolated and abstracted around fundamental business concepts, business rules are scattered over multiple systems, making code reusability difficult, if not impossible.

- **Unmanageable dependencies**: When components are not properly segregated from one another, updating or replacing a system has a snowball effect, in which changes in one portion affect all of its dependents. This becomes overwhelming quickly as businesses keep changing their requirements.

- **Legacy systems that are inflexible and slow to adjust**: It is difficult to quickly adapt a legacy system to business developments. Changes can take a long time if the system is complex and inflexible. If the technology becomes obsolete, the accumulation of core data and system dependencies over time may make replacements difficult.

Well, now that we know the need to have a proper scalable architecture for our applications, let's look at some guiding principles that will allow us to build scalable architecture next:

- **A well-tested code base**: For a scalable application architecture, tests should be a part of your development process. You should be able to write tests and reach 100% code coverage. An untested code is unscalable. Any part left untested can open doors to potential bugs. Having automated tests well integrated into the development process allows the teams to rapidly update applications without worrying about breaking any other parts.

- **Easy to debug**: Scalable code should be easy to debug. Once the code has proper test coverage, any potential bugs will fail your existing test, allowing you to figure out the exact part of the code that is causing the issue. If tests pass and the bug still exists, once the bug is fixed, you should add a related test to never allow the same error to surface again.

- **Well-documented and easy to understand**: The code base should be well-structured so that a developer at any level will be able to follow it. This can be achieved by writing consistent code and following preset standards such as using helpers such as code linters.

- **Composable**: Scalable code is also composable, meaning composed of different smaller independent modules. This allows the effective parallel development of features and modules. Each module should abstract the underlying implementation details. Each module should have a single responsibility and should be independent of other modules. They should interact with each other using the exposed abstraction.

To make an application composable and easily scalable, it's always better to separate layers domain-wise. For example, we could have the following layers in our application:

- **Infrastructure layer**: The infrastructure layer is where all the communication to external systems, such as API providers and local databases, happens.

- **Core layer**: The core layer is where the business requirements and rules are implemented. The core layer gets the information from the infrastructure layer and applies the business rules to it.

- **Application layer**: The application layer provides services to end users. This layer consists of a **user interface** (**UI**), an application state, and specific logic related to the application. This also communicates with the core layer in order to serve the users' needs.

The key thing to remember here is not to have a layer above dependent on a layer down. For instance, the infrastructure layer will be unaware of the core layer and the application layer, and the core layer will be unaware of the application layer. The dependency is only one-way. The application layer only depends on the core layer, and the core layer only depends on the infrastructure layer.

We now know why we need a scalable architecture for our application and how we can strive to make our application architecture robust and scalable. In the next section, we will look at how we can properly organize source code files and folders for better readability and support scalability better.

Organizing files and folders

Though it may seem to be a trivial task, organizing files and folders plays an important role in making code readable, maintainable, and scalable. There can be many ways to organize files and folders in Flutter. We will look at a few of them in this section. Here, we are talking about the file and folder organization for source code inside the `lib/` folder in our Flutter application.

Having proper folder organization helps us overcome the following issues:

- Being unable to find a specific file
- Writing a block of code again and again
- Mixing up the UI, business logic, and backend code
- Unlimited local variables
- Confusion when developing as a team

The first way that we can organize our files and folders is by the functionality of source code. Look at the folder structure in the following figure:

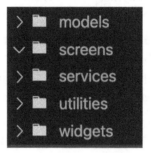

Figure 6.1 – A functionality-based folder organization

Here's what each folder comprises:

- `screens`: Contains all the screens or pages of the application. This can further be organized with subfolders such as `auth` – which contains all the authentication related to screens, such as login and sign up.
- `widgets`: Contains all the widgets required in the application.
- `utilities`: Contains utilities and functions such as API connection and constants used across the application.
- `models`: Contains data models for data used in the application.

This type of organization can be suitable for small to medium-scale applications. However, if your application has more than 20 screens and multiple external entities integrated, this can quickly get out of hand. So, let's look at another method. This is a feature-based folder organization, which is a popular folder organization method.

In the following figure, we can see the `core` and `features` folders. So, we first separate by features and then by functionality:

Figure 6.2 – A feature-based folder organization

Here, the core folder holds files and sources that are not particularly related to any feature or are shared between features. For example, the core folder can contain subfolders such as widgets, which holds common widgets shared amongst multiple features, utilities, which holds utilities used across applications, and services, which connect with the database, APIs, or other external entities, and more, such as application constants and themes.

Here, inside each feature folder, we can use the type-based organization method described previously so that each feature can now contain folders such as screens or widgets only related to that feature. This method might be overwhelming for small-scale projects with 5 to 10 features, but for an application with more than 10 features and complex organization, this can be very beneficial.

Finally, we have a combination of these methods, as well as using the layers we talked about in the previous section. We have a folder organization that we can recommend for small- to large-scale projects. Here, we first separate folders for each layer – infrastructure, core, and app. Inside each layer, we organize files and folders based on functions if we are working with small-scale applications – otherwise, we organize feature-wise if the application is medium to large scale. This is how the folder structure will look:

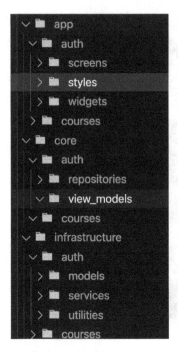

Figure 6.3 – A layer-based folder organization

Let's focus on the folders for each layer:

- `infrastructure`: This contains code that is not related to application and business logic. It is purely required for integration with external services, REST APIs, and databases. Inside this folder, we also have a `core` folder where we store the common functionalities shared across this layer. For example, we can have a `constants` file where we store REST endpoints and database-related constants. Based on the requirements, we may want to organize the contents into subfolders related to different features again. For example, this can contain `features/database` folders for local database-related codes, `features/api` for codes relating to API integration, and so on.

- `core`: This contains code that is specific to business logic. This layer provides the abstraction for the application layer, as well as interacts with the infrastructure layer to serve the application layer. We use the feature-based organization inside this folder again. For example, we can have a `features/auth` folder that stores all the authentication-related business logic files and interfaces with the infrastructure layer to provide authentication functionality.

- app: This contains code specific to the application logic. All the app's screens, widgets, themes, or anything related to the UI and user interaction are stored here and in its subfolders. In the application layer folder, we can organize the folders again either based on types (screens, widgets, and so on), as described in the first method, or features and then types, as described in the second method. We can decide this based on the scale of the application.

This allows us to clearly separate our application source code and have proper architecture. This will allow us to easily find which files and what source goes where.

Now that we know how to organize our folders depending on the size of our application, we'll see the different naming conventions that we should use in the following section.

Naming conventions

When talking about proper file and folder organization, we should also talk about naming conventions. In this subsection, we'll go through some standard naming conventions for our files, folders, and identifiers.

Use snake case for files and folder names. Snake case is a naming style where all the letters are lowercase and the words are separated by an underscore – for example, `snake_case`. In addition to that, we will use a dot (`.`) to separate the filename, type, and file extension – for example, `file_name.type.dart`. So, the name for the course page will be `courses.page.dart` and for the course view will be `courses.view.dart`. Including the type of file in the name will make it easier to find the file in a text editor or **integrated development environment** (**IDE**). Common file types that we will be using are `.page`, `.widget`, `.vm` (view model), and `.view`.

Next, we name identifiers using UpperCamelCase and lowerCamelCase. Use upper camel case for naming types and extensions. Use lower camel case for other identifiers including constants. Learn more about common standards and conventions from the official dart style guide here: `https://dart.dev/guides/language/effective-dart/style`.

Now that we know how to organize files and folders for our application, as well as some of the naming conventions, in the next section, we will look at one of the popular software architectures that will help us build scalable applications.

MVVM architecture

First, let me clarify that although we are learning about MVVM architecture here, this is not the only solution. There are many alternatives. We are just introducing one of the patterns and the whole methodology. Rest assured: you need to explore the other architectures out there, learn a few of them, and be able to select the one most suitable for you based on the context and application requirements. With that being said, let's explore the MVVM architecture itself. MVVM stands for Model, View, View Model and is one of the popular software architectures that enables application developers to separate the application logic from the business logic and allows us to build scalable applications as business needs change. Let's first learn what MVVM is and how it works.

The basic idea here with MVVM is to create a bridge between the data that you get from an external service and the application's view so that either of these can change irrespective of the other. This will make your code modular so that a view only has a dependency on the view model and not on the external data. This will allow you to easily switch between data providers without affecting your views.

The following figure shows a high-level diagram of MVVM architecture:

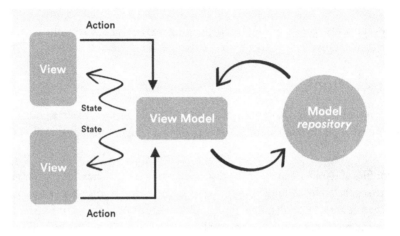

Figure 6.4 – MVVM architecture

MVVM provides a clear separation between the view, which is the presentation layer, and the model, which is the data layer. MVVM allows you to change one without affecting the other. In between the view and the model, sits the view model. The view model gets the user interactions and input from the view and sends the requests to the model. When it receives a response from the model, it notifies the view. The view model acts as a link between the view and the model, carrying all user events and returning the outcome.

There are three key benefits I can think of that we get from applying MVVM architecture:

- **Maintainability**: The presentation layer, data layer, and business logic are loosely coupled, which makes code reusable and easily maintainable. As the project grows over time, it will help you clearly distinguish between different layers and apply changes to any one layer without affecting another.

- **Testability**: The logic implemented in the view model is a lot easier to unit test and make robust without knowing about other parts.

- **Extensibility**: As the project size increases, the architecture makes sure that the application is extensible and can be built with reusable components.

Now, we know what MVVM architecture is and what benefits we get by implementing it in our application. Let's look at the three components of MVVM separately.

The model

The model represents a single source of truth that interacts with external services such as APIs and databases to provide data to the application. The model can also contain business logic and code validation. This layer interacts with the view model layer by getting data from user input or serving data to the user interactions. This layer also contains business logic and code for validation.

The view model

This is a mediator between the view and the model, hence the name view model. It accepts all the user inputs and requests appropriate data responses from the model. Once the model has the data, it returns it to the view model, and the view model notifies the view of the data. The view model may also manipulate or reformat data to make it suitable for the view to use. One view model can be used by multiple views.

The view

The view is what users see and interact with. For Flutter applications, these are the widgets that are shown on the screen. When users interact, the user events request some actions that navigate to the view model, and the rest is handled by the view model. Once the view model has data requested by the user's event, it notifies and updates the view.

Now that we know what MVVM is, its benefits, and what each layer in MVVM is responsible for, we will implement the MVVM architecture in our Flutter Academy application in the following section.

Implementing MVVM architecture

The need to implement proper architecture for the maintainability and scalability of an application is huge. Without proper architecture, it gets more and more complex to scale up the application or add new features. So, let's start refactoring our Flutter Academy application by implementing the MVVM architecture.

We will start refactoring by organizing the folders first. We will organize the folders using one of the methods we discussed previously. As this application will be a medium-sized one once we complete this book, I am choosing to use functionality-based file organization. On top of that, I will use two layers, that is, an application layer and an infrastructure layer. Follow these steps to organize the folders for your own Flutter Academy application:

1.   Inside the `lib` folder, create two folders named `app` and `infrastructure`.

2. Inside the app folder, we will again create the following folders:

    - `pages`

    - `res`

    - `routes`

    - `view_models`

    - `widgets`

3. Move all the pages, that is, `about_page.dart`, `contact_page.dart`, `course_details_page.dart`, `courses_page.dart`, `error_404_page.dart`, and `home_page.dart`, inside the `app/pages` folder.

4. Move `assets.dart` and `responsive.dart` inside the `app/res` folder.

5. Move `app_route_parser.dart` and `router_delegate.dart` inside the `app/routes` folder.

6. Move all the widgets, such as `call_to_action.dart`, `featured_section.dart`, and others, inside `app/widgets`.

    We will also rename our files using the previously discussed methods.

7. Rename `courses_page.dart` to `courses.pages.dart`. Rename the rest of the pages similarly.

8. Rename `assets.dart` and `responsive.dart` to `assets.res.dart` and `responsive.res.dart`, respectively.

9. Rename `app_route_parser.dart` as `app_route_parser.router.dart` and `router_delegate.dart` to `router_delegate.router.dart`.

10. Rename all widgets as `call_to_action.widget.dart`.

11. Now, inside `infrastructure`, create the `model` and `res` folders.

12. Move `course.dart` and `course_service.dart` to `infrastructure/model` and `infrastructure/res`, and rename them as `course.model.dart` and `course.service.dart`, respectively.

So, this is how your folder organization should look right now:

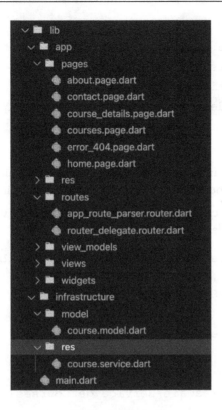

Figure 6.5 – The folder organization for Flutter Academy

Well, you have now properly organized your folders. Inside the app and infrastructure folders, you can choose to organize your folders feature-wise – for example, on top of the pages and widgets folders, you could create folders such as courses, auth, or users. So, your login.page.dart file, if it exists, would go to app/features/auth/pages/ and courses.page.dart would go to app/features/courses/pages/. Whichever pattern you follow inside the app folder, follow the same pattern inside the infrastructure folder as well. This makes it easy to follow and track your features and files.

Now, it's time to implement the MVVM pattern in our code. To build our view model, we will be using Riverpod, one of the popular state management solutions in Flutter. If you instead want to use Provider or any other state management library that you are familiar with, that's fine too. However, if you don't understand what's going on with the Riverpod-related code, no worries, we have dedicated *Chapter 8*, *State Management in Flutter*, to state management in Flutter.

So, let us first start by creating our first view model. This will be a course view model. Follow these steps to create a course view model:

1. Inside lib/app/view_models, create a file named course.vm.dart and add the following code:

```
class CourseVM {
  final Course course;
  CourseVM(this.course);
  String get title => course.title;
  String get description => course.description;
  String get image => course.image;
}
```

This will receive the `Course` model and create `CourseVM`, which we can use in our views without ever knowing about `Course` or any changes that occur in `Course`.

2. Now, create a file named `course_list.vm.dart` inside `app/view_models` and update it with the following code:

```
class CourseListVM extends StateNotifier<List<CourseVM>>
{
  final CourseService service;
  CourseListVM(this.service): super(const []) {
    fetchCourses();
  }
  Future<void> fetchCourses() async {
    final res = await service.getCourses();
    this.state = [...res.map((course) =>
      CourseVM(course))];
  }
}
final courseListVM = StateNotifierProvider<CourseListVM,
List<CourseVM>>(
    (ref) => CourseListVM(CourseService()));
```

Quite a few things are going on here. First, `CourseListVM` is a view model that returns a collection of `CourseVM`, which we are getting from `CourseService` and is defined in infrastructure. `CourseListVM` is extending `StateNotifier`, provided by Riverpod, which helps us to easily update changes and notify the listener widgets. No worries, we will go through these in detail in *Chapter 8*, *State Management in Flutter*. Finally, we create a `courseListVM` object using `StateNotifierProvider` from Riverpod and that is what we will use in our views to access the data from this view model.

To use this view model, let's create our courses view by following these steps:

1. Inside `lib/app`, create a folder named `views`.

2. Inside the `views` folder, create a folder named `courses.view.dart` and update it with the following code:

```dart
class CoursesView extends StatelessWidget {
  const CoursesView({
    Key? key,
  }) : super(key: key);

  @override
  Widget build(BuildContext context) {
    return Consumer(builder: (context, ref, child) {
      final courses = ref.watch(courseListVM);
      return ListView(
        scrollDirection: Axis.horizontal,
        children: [
          ...courses.map(
            (course) => Padding(
              padding: const EdgeInsets.only
                (top: 20),
              child: CourseCard(
                description: course.description,
                title: course.title,
                image: course.image,
                onActionPressed: () {},
              )),
          ),
        ],
      );
    });
  }
}
```

A view is simply a widget that interfaces with the view model. In the preceding view, the following piece of code interacts with our view model:

```dart
return Consumer(builder: (context, ref, child) {
  final courses = ref.watch(courseListVM);
  ....
```

The consumer again comes from Riverpod, which allows us to consume our providers. Using `ref.watch(courseListVM)`, we are listening for updates in our course list view model. Whenever any changes occur in our `courseListVM`, such as the data being received from future the widget, it will rebuild and we will be able to see the data.

Here, as you can see, we are only accessing view models from our views, and our views have no clue about the actual model or the data service. So, if the model undergoes internal changes, we change the service we receive data from; all that needs to change in our application is our view model. Again, when we update our views, we only need to change our view models. This also allows us to mock different items and test view models and views independently to build well-tested applications. Also, this is a simple view model where we are only accessing data but not interacting with it. Next, we will implement authentication, where we will not only access data but also interact with the view model using user actions such as pressing a button.

Setting up for authentication

Although we will be implementing the actual authentication using services such as Firebase or Appwrite in later chapters, right now, we will use hardcoded methods to authenticate so that we can observe the benefits of MVVM later. While using everything as it is, we can simply update the `auth` service in the infrastructure layer and properly authenticate our users without changing anything in the application layer. Let's start by creating our login page by following these steps:

1. Create a folder named `login.page.dart` inside `lib/app/pages`. Here, we will create a simple login page that will use the login view.

2. Update `login.page.dart` as follows:

    ```
    class LoginPage extends StatelessWidget {
      @override
      Widget build(BuildContext context) {
        return Scaffold(body: Center(child:
            LoginView()),);
      }
    }
    ```

    We now need to create the login view.

3. Create a folder named `login.view.dart` inside `lib/app/views`. We are going to create a simple stateful widget called `LoginView` as follows:

    ```
    class LoginView extends StatefulWidget {
      @override
      State<LoginView> createState() => _LoginViewState();
    }
    ```

```
class _LoginViewState extends State<LoginView> {
  @override
  Widget build(BuildContext context) {
    return Container();
  }
}
```

4. Next, we will initialize two text exiting controllers that will help us get a user input for the email and password:

```
class _LoginViewState extends State<LoginView> {
  TextEditingController _email = TextEditingController
      ();
  TextEditingController _password =
      TextEditingController();

  ...

  @override
  void dispose() {
    _email.dispose();
    _password.dispose();
    super.dispose();
  }
}
```

Finally, we will create a login form with an email and password field and a **Login** button.

5. Update the `build` method of `LoginView` as follows:

```
return ListView(
    shrinkWrap: true,
    padding: const EdgeInsets.all(16.0),
    children: <Widget>[
      Text(
        "Login",
        style: Theme.of(context).textTheme.
            headline4,
      ),
      const SizedBox(height: 20.0),
      TextField(
        controller: _email,
```

```
                    decoration: InputDecoration(labelText:
                        "enter email"),
                ),
                TextField(
                    controller: _password,
                    decoration: InputDecoration(labelText:
                        "enter password"),
                    obscureText: true,
                ),
                const SizedBox(height: 20.0),
                return ElevatedButton(
                    onPressed: () {},
                    child: Text("Login"),
                );
            ],
        );
```

We have now created a simple stateful widget with email and password fields along with a **Login** button. Next, we will create an auth view model that we will use with the login view to authenticate the user. Let's get started:

1. Create a folder named `auth.vm.dart` inside `lib/app/view_models`.

2. First, create an `AuthVM` class that extends `ChangeNotifier` with an `isLoggedIn` Boolean and a `String error` field as follows:

    ```
    class AuthVM extends ChangeNotifier {
        bool isLoggedIn = false;
        String error = '';
    }
    ```

    `isLoggedIn` tells us whether the user is authenticated or not. The `error` field holds any error message that might occur during login or logout.

3. Next, add a `login` method inside `AuthVM` that will receive the user's email and validate and update `isLoggedIn` as required. Add the `login` method as follows:

    ```
    bool login({required String email, required String
    password}) {
        if (email == 'test@email.com' && password ==
            'testpass') {
          error = '';
    ```

```
      isLoggedIn = true;
      notifyListeners();
      return true;
    }
    error = 'Invalid credentials';
    return false;
  }
```

We are checking the email and password against hardcoded values, as we have not integrated any authentication service right now. Once we validate that the email and password are correct, we update isLoggedIn to true and call notifyListeners().notifyListeners() is an inbuilt function provided by ChangeNotifier that helps us notify the listeners that there were changes in the values so that listeners can get the updated values. This will help us rebuild the widget whenever there is a change in data. There will be more on this in *Chapter 8, State Management in Flutter*.

4. Similarly, we will create a simple logout method inside AuthVM as follows:

```
bool logout() {
  if (!isLoggedIn) {
     error = 'Not logged in';
    notifyListeners();
     return false;
  }
  error = '';
  isLoggedIn = false;
  notifyListeners();
  return true;
}
```

Finally, we will create an instance of authVM that will be provided to the app using ChangeNotifierProvider from Riverpod.

5. Outside the AuthVM class, add the following code:

```
final authVM = ChangeNotifierProvider<AuthVM>((ref) =>
AuthVM());
```

Now that we have our auth view model, we will use it in our login view to carry out the login when a user presses the **Login** button.

In lib/app/views/login.view.dart, we will wrap ElevatedButton with a Consumer widget from Riverpod, which will help us interact with our view model as we did with the courses view model. Wrap ElevatedButton as follows:

```
Consumer(builder: (context, ref, child) {
  return ElevatedButton(
    onPressed: () {},
    child: Text("Login"),
  );
})
```

Now we can access authVM using ref provided by the Consumer widget. So, when the **Login** button is pressed, the onPressed method of the button is called. Inside that method, we will call the login function from authVm to authenticate the user as follows:

```
final result = ref.read(authVM).login(
  email: _email.text,
  password: _password.text,
)
if (result) {
  debugPrint('login successful');
} else {
  debugPrint(ref.read(authVM).error);
}
```

Here, we are calling the login method of authVM by passing the value entered by the users in the text field and printing the result of the authentication. Next, we will finally use this information to display a dashboard page instead of a home page if the user is logged in and allow them to log out. Create a folder named dashboard.page.dart inside lib/app/pages. We already have a code snippet for you for the dashboard page. Copy code snippet **6.1** from https://github.com/PacktPublishing/Taking-Flutter-to-the-Web/blob/main/snippets/chapter6.md and paste it into dashboard.page.dart.

Now that we have our dashboard page, we will update our router. We will use authVM inside the router delegate again and decide whether to show the home page or dashboard page based on whether or not the user has logged in. Let's get started:

1. Open lib/app/routes/router_delegate.router.dart and update the definition of the _getRoutes function by adding a second parameter of AuthVM:

    ```
    List<Page> _getRoutes(Uri path, AuthVM authVM) {
    ```

2. Now, we will have to update the build method using Consumer so that we can access authVM and pass it to the _getRoutes method that we are calling from the build method. So inside build, wrap the Navigator widget with Consumer:

```
    return Consumer(builder: (context, ref, child) {
        final pages = _getRoutes(_path, ref.watch(authVM));
        return Navigator(
        ...
        );
    });
```

3. Now that we have access to `authVM`, in the `_getRoutes` function, the first instance where we create a `pages` array containing a home page, update as follows to add either the home page or the dashboard page based on whether the user is authenticated or not:

```
    Final pages = <Page>[];
        if (authVM.isLoggedIn) {
          pages.add(MaterialPage(child: DashboardPage(),
              key: ValueKey('home')));
        } else {
          pages.add(MaterialPage(child: HomePage(), key:
              ValueKey('home')));
        }
```

So, if the user is authenticated, we will show the dashboard page – if not, we will show the home page as the default page.

4. Next, we will also update the `/login` case to redirect back to the root route if already authenticated:

```
    case 'login':
        if (authVM.isLoggedIn) {
          go('/');
          break;
        }
        pages.add(MaterialPage(
          key: ValueKey('login'),
          child: LoginPage(),
        ));
        break;
```

We simply check whether `authVM.isLoggedIn` is `true` and redirect the user back to the root route if that's true.

Now, if you run your application and first go to /login, you will see the login form. Next, if you type the correct email and password, you will be redirected to the root route and you can see the dashboard page instead of the home page. Now, if you try to go to /login again, you will be redirected to the dashboard page. So, here we use the auth view model to authenticate the user. Later, when we implement the actual authentication using services such as Firebase or Appwrite, we will not have to change anything in our views – we only change our view models. That's the beauty of MVVM.

Finally, we also want users to be able to log out so, in lib/app/pages/dashboard.page.dart, we will wrap the NavigationRail widget with a Consumer widget and use authVM to log the user out when they select the **Log out** option. So, update NavigationRail on the dashboard page as follows:

```
Consumer(builder: (context, ref, child) {
  return NavigationRail(

    ...

    onDestinationSelected: (dest) {
      if (dest == 2) {
        ref.read(authVM).logout();
      }
    },
  );
}),
```

Once we wrap with Consumer, we just need to update the onDestinationSelected function to call authVM's logout method when the selected destination is 2, as in the **Log out** button for us. Now if you press the **Log out** button, you will be redirected back to the home page and you should be able to access the login page by going to /login.

That's all for this section. We have seen just how easy it is to implement the MVVM pattern in Flutter to make our applications robust and scalable.

Summary

In this chapter, we looked at the importance of having proper architecture in our applications and learned about the various issues it can solve. We also looked at different ways of organizing our source files and the benefits of each way. Then, we learned about various naming conventions that we can follow in our application to make it easier for everyone to work with it. Finally, we learned about MVVM architecture and implemented it in our application.

In the next chapter, we will learn about data persistence, and how we can save things such as user preferences and other settings so that each time the application restarts, users will not lose their data.

Part 3: Advanced Concepts

This part will focus on advanced concepts regarding persisting data, integrating external services, and more.

This part comprises the following chapters:

7
Implementing Persistence

In the previous chapter, we learned about scalable architecture and different ways to organize files and folders. We also learned about the MVVM pattern and updated our application using this pattern. In this chapter, we will learn about data persistence. The persisted data is available to the user across their sessions even when they terminate the application or restart their device. We will see what different tools and techniques there are to persist user data over various sessions and how to use them to persist any kind of user data. We will also look at various plugins that help us persist data and will implement one of them in our application.

By the end of this chapter, you will understand the different tools and techniques available for persisting data in Flutter. You will have learned about each of them and how to use them. You will also know how to persist user data over different sessions and restore available data whenever a user opens the application.

In this chapter, we will cover the following topics:

- The need to persist data
- The tools and techniques available for persisting data
- Persisting data with local storage using the `shared_preferences` plugin
- Persisting data using HiveDB

Technical requirements

The technical requirements for this chapter are as follows:

- Flutter version 3.0 or later installed and running
- Visual Studio Code or Android Studio
- Google Chrome browser

You can download the code samples for this chapter from the book's official GitHub repository at `https://github.com/PacktPublishing/Taking-Flutter-to-the-Web`. The starter code for this chapter can be found inside the `chapter7_start` folder.

The need to persist data

When building any kind of mobile application, we are trying to solve users' problems, provide them with an easy way to achieve certain tasks, or entertain them. Whatever we are doing, there's data involved. Whether it's a simple to-do app or a complex game, the application needs this data to function properly. So, this data has to be persisted throughout the life cycle of an application. When talking about data persistence, we can think of two types. The first is remote data persistence, where we store data in a remote server and access it via HTTP requests. The second is offline or local data persistence, where we save the data locally on a device and access it without the internet. However, local data is available only on that device; if the device is reset or the user changes the device, this data is not available. Even when remote data is involved, we sometimes persist it locally to make the application responsive and performant.

We will learn about remote data persistence in later chapters. In this chapter, we will be focusing on local data persistence, looking at the different methods available for persisting data, and implementing a few of them. We now know why we need to persist user data. We will look at the tools and techniques available to persist data locally in the next section.

The tools and techniques available for data persistence

As we said previously, persisting data is vital to an application's proper functioning. So, let's look at different ways in which we can persist data in Flutter applications. Talking about persistence techniques, there are three broad types: persisting a key-value pair, saving data in files, and finally, saving data using SQL databases. Persisting key-value pairs is the simplest method and can be used for simple data, mostly primitive types. By using key-value pairs, we can persist simple user settings, such as the chosen theme, or things such as the offline cache of an API query where the URL will be the key. Using files, we can write any kind of data; however, managing those ourselves is a challenge. Both of these methods, using key-value pairs or files, don't provide us with advanced features, such as searching and filtering. We can only save and get data. All the remaining operations can happen in memory.

Using a SQL database (mostly SQLite in local persistence), we can persist complex data, as well as query it effectively. However, using a SQL database is not yet supported in Flutter web, so we will not be talking about it much here. If you have previously used SQL databases, such as MySQL, MS SQL, or Postgres, it's all the same concept.

In Flutter, there are various official and community-developed plugins and packages to help us persist data. Let's look at some of them:

- **sqflite** (`https://pub.dev/packages/sqflite`): This package is for directly working with SQLite databases, which we can use to store complex relational data and perform queries on them. However, this is not yet supported on Flutter web so we will not be discussing it further here. You can learn more about it from the preceding link, and if you know SQL, then this should be fairly simple for you to implement.

- **Drift** (`https://pub.dev/packages/drift`): Drift is a reactive data persistence library built on top of SQLite. This does support the web using `sql.js`; however, the web support is still experimental. So, we will be skipping this as well.

- **SharedPreferences** (`https://pub.dev/packages/shared_preferences`): This is one of the most popular plugins for persisting key-value pairs and is an official plugin supported by Flutter. The name comes from Android, where shared preference is a part of the official Android spec and is widely used in Android applications. On iOS, it uses NSUserDefaults to achieve this. On the web, it uses local storage. As suggested in the package's README file, this is not reliable for storing critical data. However, we can use it for saving simple data, such as user personalization and offline cache. We will look into this in detail in the next sections.

- **Hive** (`https://pub.dev/packages/hive`): Hive is another popular package for key-value data persistence in Flutter. It is written in pure Dart, and it supports all Dart and Flutter platforms. We will look into this in detail in the following sections.

You can find even more options if you go to `https://pub.dev`; however, these are some of the popular options. Well, now that we know about the different tools available for persisting data, in the next sections, we will further explore some of the tools and implement them in our application.

Persisting data with local storage using the shared_ preferences plugin

The `shared_preferences` plugin provides one of the easiest ways to persist data in Flutter applications. It can be used to store simple key-value pairs and is suitable for storing things such as application settings and offline cache. This plugin uses underlying `SharedPreferences` in Android, NSUserDefaults in iOS, and `localStorage` on the web to achieve key-value persistence. You can learn more about the web's `localStorage` and its API from the official specification: `https://developer.mozilla.org/en-US/docs/Learn/JavaScript/Client-side_web_APIs/Client-side_storage`. Using SharedPreferences, we will build a simple theme switcher, where users will be able to choose whether the theme is dark or light, which will persist the next time they come back.

Creating a theme mode service

To create a theme mode using SharedPreferences, follow these steps:

1. We will start by adding it as a dependency in our application. Open the `pubspec.yaml` file and under the dependencies, add `shared_preferences: ^2.0.11`.

You can use the latest version by going to `https://pub.dev/packages/shared_preferences`. Once you have added the `shared_preferences` package, you can write a service that will handle saving and getting values using SharedPreferences.

2. Let's add a file named `theme_mode.service.dart` inside `lib/infrastructure/res`.

3. Inside this file, import the `shared_preferences` package and add the following class:

```dart
import 'package:shared_preferences/shared_preferences.
    dart';


class ThemeModeService {
  static ThemeModeService? _instance;


  ThemeModeService._();


  static ThemeModeService get instance {
    if (_instance == null) {
      _instance = ThemeModeService._();
    }
    return _instance!;
  }
}
```

Here, we are defining a `ThemeModeService` class and making it a singleton by making the constructor private and providing an instance getter to always get a single instance.

4. Now, inside the class, first, let's add a method that will save a Boolean value signifying whether or not to use dark mode:

```dart
Future<bool> saveThemeMode(bool darkMode) async {
  try {
    SharedPreferences prefs = await
        SharedPreferences.getInstance();
    prefs.setBool("darkMode", darkMode);
    return true;
  } catch € {
    return false;
  }
}
```

Here, we have an asynchronous method called saveThemeMode, where we pass a Boolean parameter. This saves the Boolean value under the darkMode key. We need the key later to fetch the value from storage.

5.   Next, let's add a method to get the value from storage. It's almost the same as the previous one:

```
Future<bool> isDarkMode() async {
  try {
    SharedPreferences prefs = await
        SharedPreferences.getInstance();
    return prefs.getBool("darkMode") ?? false;
  } catch €{
    return false;
  }
}
```

Here, we are getting the value from storage using the same key. It might return null, so we have set the default value to false.

Now, our theme mode service is complete. Next, let's write our view model for the theme mode.

Creating a theme mode view model

Follow these steps to create a theme mode view model:

1.   First, create the lib/app/view_models/theme_mode.vm.dart file and create a class that extends ChangeNotifier, as follows:

```
class ThemeModeVM extends ChangeNotifier {
  ThemeModeService _themeModeService;
  ThemeMode _themeMode = ThemeMode.dark;
  ThemeModeVM(this._themeModeService) {
    loadThemeMode();
  }

  ThemeMode get themeMode => _themeMode;
}
```

The theme mode view model requires the theme mode service that we defined previously to handle the persisting and loading of the theme mode selected by the user in storage.

2.   Next, we will create a provider that will provide the view model (no worries, you will learn more about providers in the next chapter). Below the ThemeModeVM class in lib/app/ view_models/theme_mode.vm.dart, add the following code:

```
final themeModeProvider =
    ChangeNotifierProvider((_) => ThemeModeVM
        (ThemeModeService.instance));
```

3. Next, let's add a function that will load the theme mode from storage:

```
Future<void> loadThemeMode() async {
    final darkTheme = await _themeModeService.
        isDarkMode();
    _themeMode = darkTheme ? ThemeMode.dark :
        ThemeMode.light;
    notifyListeners();
}
```

Here, we are calling the theme mode service to load if dark mode is selected, as well as setting the local _themeMode variable and calling notifyListeners() so that all the UIs that are listening to this view model will update.

Finally, we want a function to change the theme mode when the user switches.

4. Let's add the toggleThemeMode method, which will toggle the theme mode and save the value using the theme mode service. This method will also update the local state and notify listeners, as follows:

```
Future<void> toggleThemeMode() async {
    _themeMode = _themeMode == ThemeMode.light ?
        ThemeMode.dark : ThemeMode.light;
    await _themeModeService.saveThemeMode(themeMode ==
        ThemeMode.dark);
    notifyListeners();
}
```

Well, now that we have our view model, let's use it in our main app widget to provide the theme mode to the MaterialApp widget.

Integrating our view model in a widget

Follow these steps to integrate the view model in the MaterialApp widget:

1. Open the main.dart file and first, wrap the MaterialApp widget with Consumer and AnimatedBuilder. We want to change the theme to be animated, so we are wrapping it with AnimatedBuilder as well. The consumer will get themeModeVM, which will be used as the animation value for AnimatedBuilder. The build method of the MyApp widget should look as follows:

```
return Consumer(
  builder: (context, ref, child) {
    final themeModeVM = ref.watch(themeModeProvider);
    return AnimatedBuilder(
      animation: themeModeVM,
      builder: (context, child) {
        return MaterialApp.router(
          //...existing properties
          themeMode: themeModeVM.themeMode,
        );
      }
    );
  }
);
```

2. For the `MaterialApp` widget, keep all the existing properties and set the `themeMode` property to `themeModeVM.themeMode` so that `MaterialApp` will load with the correct theme.

Finally, we need to allow users to switch the theme. So, in the top navigation bar, we will add a button that will allow users to switch between the dark and light themes.

3. Open `lib/app/widgets/top_nav.widget.dart` and after the contact button, let's add the following button, which will allow switching the theme mode:

```
Consumer(builder: (context, ref, child) {
  final themeModeVM = ref.watch(themeModeProvider);
  return TextButton(
    child: Text(themeModeVM.themeMode ==
        ThemeMode.dark
      ? "Light Theme"
      : "Dark Theme"),
    style: TextButton.styleFrom(
      primary: Colors.white,
    ),
    onPressed: () {
      themeModeVM.toggleThemeMode();
    },
  );
})
```

Here, we have a Consumer widget that allows us to consume our view model and uses the view model to update the theme. Whenever the user switches the theme, the view model notifies the listener, and our MyApp widget, which is also consuming this view model, receives the update and rebuilds to load with the proper theme selected.

4. To verify that our code is also persisting this value, you can toggle the theme and reload the web page. You should see the theme you last selected, not the one we assign by default.

5. Also, to confirm the value is being saved in the local storage, you can open the developer tools in the Google Chrome browser and go to the **Application** tab.

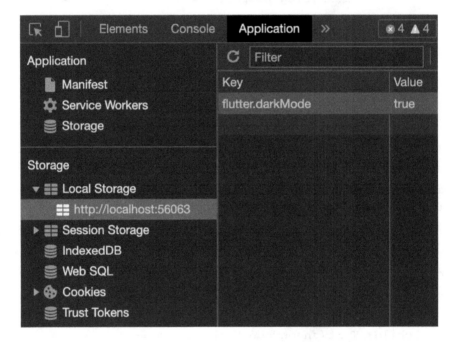

Figure 7.1 – The persisted theme mode value

In the local storage, you should be able to see the value saved there and the value here changes as you toggle the theme mode.

That's it. Using shared_preferences, we were able to build a simple persisted user setting. The user-chosen theme mode is stored in the browser's local storage so whenever users come back to our application, it will remember that the user had chosen dark mode and would not need to apply the dark mode theme again.

In the next section, we will learn to persist data using HiveDB, which has a bit more functionality than shared_preferences and is suitable to store more complex data.

Persisting data using HiveDB

HiveDB is one of the other popular choices for persisting data in Flutter. As it's written completely in Dart, it supports the web natively. In this section, using HiveDB, we will build a watchlist feature. We will allow users to add any course of their choice to the watchlist and we will add a watchlist section where users can view a list of their watched courses. We are using HiveDB for this feature as it is designed to handle a bit more complex data than SharedPreferences. Also, HiveDB has an easier API to implement. We can use HiveDB to store and load the theme mode that we implemented in the previous section as well without any issues.

Setting Up HiveDB

Follow these steps to set up HiveDB:

1. We will start by adding the dependency to our `pubspec.yaml` file. To use HiveDB with Flutter, we will add the `hive` and `hive_flutter` packages as the dependency. Under dependencies in `pubspec.yaml`, add the following:

    ```
    dependencies:
        hive: ^2.0.5
        hive_flutter: ^1.1.0
    ```

    Once we've added the dependency, before we can use HiveDB, we need to initialize it in our `main` method.

2. Open `main.dart` and update the `main` method, as follows:

    ```
    void main() async {
      WidgetsFlutterBinding.ensureInitialized();
      await Hive.initFlutter();
      runApp(ProviderScope(child: MyApp()));
    }
    ```

    Here, we are making the `main` method asynchronous. Once we do that, we need to call `WidgetsFlutterBinding.ensureInitialized` to make sure Flutter works properly. Then, we call `Hive.initFlutter()` so that the HiveDB is initialized properly and ready to use.

Next, we can move on to writing the watchlist service that will help us put and get data from HiveDB.

Writing a watchlist service

Follow these steps to write a watchlist service:

1. Create a `watchlist.service.dart` file inside `lib/infrastructure/res` and write a class-preparing singleton as we did with the theme mode service in the previous section:

    ```
    class WatchlistService {
      static WatchlistService? _instance;
      final String boxName = 'watchlist';
      WatchlistService._();

      static WatchlistService get instance {
        if (_instance == null) {
          _instance = WatchlistService._();
        }
        return _instance!;
      }
    }
    ```

    HiveDB provides named boxes to store data. Here, we have a name for our watchlist box that we will use in the next methods to store and retrieve data.

2. Let's now add a method that will save a course in the watchlist:

    ```
    Future<void> addToWatchlist(int id, Map<String,
        dynamic> course) async {
      final box = await Hive.openBox<Map<String,
          dynamic>>(boxName);
      if (box.get(id) == null) {
        box.put(id, course);
      }
    }
    ```

    This is how simple it is to work with HiveDB. First, we called `Hive.openBox` and set the type of data to be stored in the box; for us, it's `Map<String, dynamic>`. Then, we provided a name for the box. After that, we used the `put` method from the opened box to put a key and a value inside the box. Here, the key is the ID for the course and the value is the map data for the course.

3. Similarly, make a method to remove a course from the watchlist:

    ```
    Future<void> removeFromWatchlist(int id) async {
    ```

```
    final box = await Hive.openBox(boxName);
    box.delete(id);
}
```

Similar to what we did in the `addToWatchlist` method previously, we opened the box and used the `delete` method provided by the box.

4. Next, let's also write a method that will return whether or not the course is in the watchlist:

```
Future<bool> isInWatchlist(int id) async {
    final box = await Hive.openBox(boxName);
    return box.containsKey(id);
}
```

Again, we opened the box and used the `containsKey` method to check whether or not the ID is present in the box, simple as that.

5. Finally, we will write a method that returns a list of courses in the watchlist, as follows:

```
Future<Iterable<Map<String, dynamic>>> getWatchlist()
async {
    Final box = await Hive.openBox<Map<String,
        dynamic>>(boxName);
    return box.values;
}
```

Finally, HiveDB provides us with a listenable that will allow us to listen to changes to the values in the box. That is what we are returning from the `getWatchlist` method so that we can build our widget that will list courses in the watchlist.

As we are following the MVVM pattern, we will next create a watchlist view model.

Creating a watchlist view model

Follow these steps to create a watchlist view model:

1. Inside `lib/app/view-models`, create a file named `watchlist.vm.dart`. Inside this, we will create our view model class, as follows:

```
class WatchlistVM extends StateNotifier<List<CourseVM>> {
    WatchlistService _watchlistService;
}
Final watchlistVM = StateNotifierProvider<WatchlistVM,
        List<CourseVM>>((_) => WatchlistVM
(WatchlistService.instance));
```

Our watchlist view model will use the watchlist service we created previously to handle saving and loading data from the watchlist. Here, we are also creating `watchlistProvider` to use this view model in our views.

2. Next, we will add the methods to retrieve, add, and remove data in the watchlist, as follows:

```
WatchlistVM(this._watchlistService) : super(const []) {
  getWatchlist();
}


Future<void> addToWatchlist(int id, Course course)
async {
    await _watchlistService.addToWatchlist(id,
        course.toMap());
    this.state = [CourseVM(course), ...this.state];
}


Future<void> removeFromWatchlist(int id) async {
    await _watchlistService.removeFromWatchlist(id);
    this.state = this.state.where((course) =>
        course.course.id != id).toList();
}


Future<void> getWatchlist() async {
    final watchlist = await
    _   watchlistService.getWatchlist();
    this.state = [...watchlist.map((e) =>
        CourseVM(Course.fromMap(e)))];
}


bool isInWatchlist(int id) {
    return this.state.where((element) =>
        element.course.id == id).length != 0;
}
```

Here, everything is already handled by our watchlist service. In each method, we are calling the related method from the watchlist service.

Next, we will add a button to our course card and course details page that will allow users to add a course to their watchlist.

Integrating a view model into views

Follow these steps to integrate the view model into views:

1. Open `lib/app/widgets/course_card.widget.dart` and after the description on line 55, add the following button:

```
Center(
  child: IconButton(
    onPressed: () {},
    icon: Icon(Icons.add),
  ),
),
```

2. Next, we will wrap the icon button with a consumer to read our view model to add or remove items from the watchlist, as follows:

```
Center(
  child: Consumer(builder: (context, ref, child) {
    final isInWatchlist =
        ref.watch(watchlistVM.notifier).
            isInWatchlist(id);
    return IconButton(
      onPressed: () {
        if (isInWatchlist) {
          ref.read(watchlistVM.notifier).
              removeFromWatchlist(id);
        } else {
          ref.read(watchlistVM.notifier).addToWatchlist(
              id,
              Course(
                id: id,
                title: title,
                description: description,
                image: image,
              ),
          );
        }
      },
```

```
        icon: Icon(isInWatchlist ? Icons.clear :
            Icons.add),
      );
    }),
  ),
),
```

The course card now shows the button to add a course to the watchlist.

Taking Flutter for Web

Flutter web is stable. But **There** are no proper courses. focused on Flutter web. So, In this course, we will learn what Flutter for Web is good for and w...

+

Figure 7.2 – A course card with an "add to watchlist" button

Now that we have a way to add a course to the watchlist, we will set up navigation and create a page to view the courses added to the watchlist.

3. First, in `top_nav.widget.dart`, we will add a new button that will navigate us to the `/watchlist` route. After the about button, add the following button:

```
TextButton(
  onPressed: () {
    routerDelegate.go('/watchlist');
  },
  child: Text("Watchlist"),
  style: TextButton.styleFrom(
    primary: Colors.white,
  ),
),
```

This is how the top navigation bar looks:

Flutter Academy	Home Courses About Watchlist Login Contact Dark Theme

Figure 7.3 – The top navigation bar

Next, we will create a watchlist page.

Creating a watchlist page

Follow these steps to create a watchlist page:

1. Create a new file called `lib/app/pages/watchlist.page.dart`. Inside this, we will create a widget that will use the watchlist view model and list the courses added to the watchlist. First, we will create a basic layout reusing the widgets we already have:

    ```
    import 'package:flutter/material.dart';
    import '../res/responsive.res.dart';
    import '../widgets/drawer_nav.widget.dart';
    import '../widgets/top_nav.widget.dart';

    class WatchlistPage extends StatelessWidget {
      @override
      Widget build(BuildContext context) {
        return Scaffold(
          body: Column(
            children: <Widget>[
              TopNav(),
              Expanded(
                child: Container()
              ),
            ],
          ),
          drawer: MediaQuery.of(context).size.width >
              ScreenSizes.md
              ? null
              : DrawerNav(),
        );
      }
    }
    ```

2. Next, below `TopNav`, we will modify the child of the `Expanded` widget, which is where we will list courses from the watchlist. So, there, we will start by adding a consumer, as follows:

```
Expanded(
  child: Consumer(
    builder: ((context, ref, child) {
      final List<CourseVM> courses = ref.watch
        (watchlistVM);
    }),
  ),
),
```

Well, now we have a list of courses from the watchlist using `watchlistVM`.

3. Next, we will add a grid view to display the list of courses in the watchlist, as follows:

```
builder: ((context, ref, child) {
  final width = MediaQuery.of(context).size.width;
  final List<CourseVM> courses = ref.watch
    (watchlistVM);
  return GridView.builder(
    gridDelegate:
SliverGridDelegateWithFixedCrossAxisCount(
      crossAxisCount: crossAxisCount: width >
        ScreenSizes.xl ? 4 : width > ScreenSizes.lg
          ? 3 : width > ScreenSizes.md ? 2 : 1,
    ),
    itemBuilder: (context, index) {
      final course = courses[index];
      return CourseCard(
        id: course.course.id,
        image: course.image,
        title: course.title,
        onActionPressed: () {},
        description: course.description);
    },
    itemCount: courses.length,
  );
}),
```

Now that we have completed our watchlist page, the next step is to add its route. We want to display the watchlist page under the /watchlist route.

4. Add a new case in the lib/app/routes/router_delegate.router.dart file's _getRoutes function, as follows:

```
case 'watchlist':
  pages.add(MaterialPage(
    child: WatchlistPage(),
    key: ValueKey('watchlist'),
  ));
  break;
```

Well, we now have everything set. If we now run our project using flutter run -d chrome, we should be able to see an add button in the course card, and clicking it will add the course to the watchlist. Then, we should be able to navigate to /watchlist to view the list of courses that we added to the watchlist, as follows:

Figure 7.4 – The watchlist page

With that, we are done. Using HiveDB, we implemented a watchlist where users could bookmark the courses of their choice. The data will remain whenever they come back to the application again.

Summary

In this chapter, we looked at the different ways to persist user data locally. We learned why data persistence is required and what options are available in Flutter to persist data. Finally, we implemented SharedPreferences and HiveDB in our application to persist data. We started by creating a service class that deals with `shared_preferences` or HiveDB to save and retrieve data using the plugins. Then we followed the MVVM pattern that we discussed in *Chapter 6, Architecting and Organizing*, to create a view model and implement the functionality to save data using the service class we defined. This way, it keeps code well organized and maintainable.

In the next chapter, we will learn about state management, which is one of the most important techniques as your application state starts getting complex. We will learn about different state management options and implement one of them. See you in the next chapter!

8
State Management in Flutter

In the previous chapter, we learned about data persistence, the different tools and techniques that are available to persist a user's data over various sessions, and how to use them to persist any kind of user data. We also looked at various plugins that help us to persist data and implemented some of them in our application. In this chapter, we are going to cover another important topic for application developers – state management. We will cover what it is, what options are available, and how to implement one of them in our application.

By the end of this chapter, you will understand the basics of state management. You will also know about the various tools and techniques available for state management in Flutter and ultimately learn to use one of them. The concepts you learn here should help you implement any of your preferred state management options by following their guidelines.

In this chapter, we will cover the following topics:

- The basics of state management
- The options available for state management
- An introduction to Riverpod
- Implementing Riverpod in our app

Technical requirements

The technical requirements for this chapter are as follows:

- Flutter version 3.0 or later, installed and running
- Visual Studio Code or Android Studio
- Google Chrome browser

You can download the code samples for this chapter and other chapters from the book's official GitHub repository at `https://github.com/PacktPublishing/Taking-Flutter-to-the-Web`. The starter code for this chapter can be found inside the `chapter8_start` folder.

Understanding the basics of state management

Before we begin with the basics of state management, there is one thing we need to understand, and that is that Flutter is a declarative framework. Flutter builds the user interface based on the current state, in other words, *UI = f(state)*, as in, UI is the function of the state. We do not imperatively change the UI – for example, in Android, we are used to `textView.setText()`. However, in Flutter, we just change the value that the `Text` widget is bound to, and Flutter rebuilds the widget to reflect the change in value.

Now that we understand the declarative framework, let's begin with the basics of state management. For that, we first need to understand what the state of an application is. If we are talking in the broadest possible sense, an application state is anything that is in memory while the application is running, including assets, variables, fonts, and the animation state, and most of these things are managed by Flutter internally, so we don't need to worry about them. The only application state that we are concerned about is the data that we need to build the UI at any particular instant. And that is what we need to manage.

The state that we need to manage is again divided into two types – the ephemeral state and the app state.

An **ephemeral state**, or the UI or local state, is one that is mostly contained in a single widget – for example, the progress regarding an animation, the on/off state of a switch, or a currently selected item in `BottomNavigationBar`. The point here is that this state is not shared between the other parts of the widget tree. This is easy to handle and takes little to no effort. For this kind of state, Flutter provides and recommends the use of a **StatefulWidget** class. A `StatefulWidget` class can manage its own state and whenever we want to change any value in the state, we call `setState()` and the widget rebuilds itself.

Finally, a state that we want to share across different parts of our application and keep between user sessions is called the **app state** – for example, user preferences, login info, and a shopping cart in an e-commerce app. The app state varies between applications and can comprise either simple or complex data based on the requirements of our application. To manage the state, we will need to do research into the different options available and choose the one that is most well suited to our application and our team.

Before concluding this section, there is one more thing I wish to clarify; there is no predefined or clear rule through which we can decide that something is an app state while another thing is an ephemeral state. In fact, we could just use `StatefulWidget` with `setState()` to manage the state of our entire application; however, based on the nature and complexity of the application, this might become difficult to manage.

Now that you have an understanding of what the application state is, let's understand the basic principle of state management and how it works in a reactive framework such as Flutter. First, an application state is just data that is required by the application, as already discussed previously. Hence, any state management solution must provide a way to store any type of data. A state can change, so the state management solution has to provide an easy way to change the data – in other words, the state.

The change may occur based on user actions, network activity, or any other events. And in a reactive framework, state management should also provide a way for the UI to listen to changes in the state so that whenever the state changes, the UI can update itself. If we look at a basic state management solution, `StatefulWidget` provided by Flutter, we can see these state management solution properties in action.

For example, in `StatefulWidget`, we can store any type of data supported in a variable. The UI in the widget is already listening to the changes to any data in that widget itself and `setState()` is the function used to provide notification of the state change. We can call `setState()` based on user action, network event, or any other supported events. This is how basic state management works, irrespective of which state management solution you use. Now that we have a basic understanding of what state is and when we require a state management solution, next we will explore the various options available for managing our application's state in Flutter.

The options available for state management

State management is vital for any decent, production-grade application to make it scalable and easy to maintain. However, there is no one state management solution for everyone. There isn't a single official, best state management solution, even though the Flutter team does recommend several solutions. However, this choice depends on your requirements and your team's preferences. In this section, we will look at the various options available. However, keep in mind that these are not the only options; more are available, and by the time this book is released and gets to you, further solutions may have popped up in the Flutter community. Also, keep in mind that people using any state management solution may build their own state management logic or implementation on top of the existing ones. Let's look at a few options available, which are also very popular.

Provider

Provider (`https://pub.dev/packages/provider`), one of the popular and recommended approaches for state management in Flutter, is simple and easy to get started, yet very powerful. It is one of the recommended state management solutions used widely in the community. You will find lots of resources regarding using Provider to manage state in Flutter.

Riverpod

Riverpod (`https://pub.dev/packages/riverpod`) is another state management tool written from the ground up to fulfill the shortcomings of Provider and has been developed by the same developer. It is very similar to Provider, although not dependent on Flutter. We can use Riverpod on our Dart-only projects. This is the new solution recommended by the developer of Provider himself and this is also the solution that we will be learning about in this chapter.

Redux

Redux (`https://pub.dev/packages/redux`) is a Dart implementation of the JavaScript version of the widely popular state management solution of the same name. Redux is a predictable state container for Dart and Flutter apps. It uses a combination of action and reducers to manage the application's state.

BLoC

BLoC (`https://pub.dev/packages/bloc`) is a state management library that helps implement **Business Logic Component (BLoC)** design patterns. It is simple and lightweight, highly testable, and can be used in Dart and Flutter projects.

MobX

MobX (`https://pub.dev/packages/mobx`) is a library for reactively managing the state of your applications. It uses observables, actions, and reactions to build a state management solution for Dart and Flutter projects.

To learn more about each state management solution, you need to refer to the official documentation of each package. You will also find lots of tutorials and videos regarding these solutions published by the community. However, you need to understand that the basic principles stay the same.

Now that we know various options available for state management, in the following section, we will learn about Riverpod as a state management solution.

An introduction to Riverpod

Riverpod is an excellent state management solution for Dart and Flutter applications. It was written from the ground up by the developer of Provider, taking everything learned from the Provider package to overcome its shortcomings. Unlike Provider, Riverpod can also be used on Dart-only projects without Flutter.

In this section, we will look at how to install Riverpod and use it in our application. We will also better understand the view models that we wrote about in *Chapter 6, Architecting and Organizing*, as they already use Riverpod.

Install the latest version of Riverpod from `pub.dev` (at the time of writing, it's 1.0.3). Follow these steps to install Riverpod:

1. Add the following under the dependencies of your `pubspec.yaml` file:

    ```
    flutter_Riverpod: ^1.0.3
    ```

2. Then, run the following command:

```
flutter pub get
```

This particular package in the Riverpod family is optimized to be used with Flutter and provides some wrapper widgets that make accessing states in widgets easy. There are also other packages. You can learn more about which package to choose by going to `https://Riverpod.dev/docs/getting_started`. After that, we need to wrap our root widget with `ProviderScope` from the `flutter riverpod` package.

3. Modify the main function in `main.dart` as follows:

```
void main() async {
  runApp(ProviderScope(child: MyApp()));
}
```

Now that we have installed Riverpod and wrapped our root widget with `ProviderScope`, we can start using Riverpod. However, before we do that, let's get to know some of the components of Riverpod, which will help us to understand the implementation a bit better.

Understanding Providers

A provider is simply an object that encapsulates a piece of a state and allows us to listen to it. It allows us to easily listen to states in multiple locations. It provides internal methods to combine the current state we are defining with others. It also increases the testability of our application. There are different types of providers available in the package for different purposes. Here is a list of providers with their descriptions:

*   **Provider**: This is the simplest of all providers and can be used for values of any type. This is only useful when the state doesn't change.

*   **StateProvider**: This, too, is simple, as with the basic provider, although it provides a method to modify the state. It is a simple form of `StateNotifierProvider` and is designed for simple use cases.

*   **StateNotifierProvider**: This is the recommended way to manage states in Riverpod. It is a provider for `StateNotifier` that comes from the `state_notifier` package, which is already available with `flutter_riverpod` and doesn't need to be installed separately. `StateNotifier` makes it easy to expose an immutable state and also encapsulates the methods for modifying the state – in other words, the business logic.

*   **FutureProvider**: This is equivalent to `Provider`, but for asynchronous code. It assists with the straightforward caching of async operations and provides nice and easy methods for error handling, while also allowing to combine multiple asynchronous values.

- **StreamProvider**: This is similar to `FutureProvider`, but for streams. Using `StreamProvider` has a few benefits over dealing with `StreamBuilder` and streams directly. `StreamProvider` caches the latest emitted value, allows other providers to listen to it using ref, and makes error handling easy, for example.

Next, we need to learn two important concepts – how to create a provider and how to access a provider. Once we know these two things, we can start implementing Riverpod in our application.

Creating a provider

In Riverpod, creating any kind of provider follows this same pattern:

```
final myProvider = Provider((ref) {
  return MyValue();
});
```

This code snippet consists of three components:

- `final myProvider`: The declaration of a variable. We use this variable to read the state of the provider. Providers should always be final.

- `Provider`: The provider that we decided to use. This could also be any of the other providers we discussed previously.

- An anonymous function that creates the shared state. The function will always receive an object of `WidgetRef`. This object provides access to other providers and allows us to combine or manipulate the current provider based on other providers, the life cycle methods of the providers, and more.

A function is a different object depending on the provider used. For example, a `StreamProvider`'s callback should always return a stream, whereas a simple provider can create and return any object.

Now that we know how to create a provider, next, we will learn how to read a provider to receive whatever value the provider is holding.

Reading a provider

We need access to a `WidgetRef` object to read any provider and there are two places where we can get access to a `WidgetRef` object – first, inside another provider, where we are already provided with the reference object in the callback function, and second, inside a widget tree using various widgets.

The `WidgetRef` object provides three methods for reading a provider – `watch()`, `listen()`, and `read()`. The `watch()` method is used to obtain the value of a provider and subscribe to the changes. Using this, whenever the state changes, the widget will rebuild. The `listen()` method allows us to add a listener to a provider object so that we can act in some way whenever the provider changes – for example, showing and hiding a dialog. Finally, we use the `read()` method to obtain the value of the provider but ignore the provider changes. This is particularly useful when we are obtaining a provider value inside click actions. It is officially recommended to use the `watch()` method whenever possible. This will make our application reactive, declarative, and ultimately, more maintainable.

First, get `WidgetRef` inside a provider. All providers receive a `WidgetRef` object as a parameter when the provider is created – for example, see the following:

```
final provider = Provider((ref) {
  // we can then use ref to access other providers
  final repository = ref.watch(repositoryProvider);

  return SomeValue(repository);
})
```

This means we can also pass this `ref` object to any value we create inside the provider and use `ref` in those values to access other providers.

Next, we need to get the `WidgetRef` object inside the widget in the `build` method. We can do this in two ways. One, we can extend our `StatelessWidget` with `ConsumerWidget` and `StatefulWidget`, and `State` with `ConsumerStatefulWidget` and `ConsumerState`, respectively – for example, see the following:

```
class CourseList extends ConsumerWidget {
  const CourseList({Key? key}) : super(key: key);
  @override
  Widget build(BuildContext context, WidgetRef ref) {
    ...
  }
}
```

We can also choose to wrap a specific widget with a `Consumer` widget instead, which provides a `WidgetRef` object as follows:

```
return Consumer(builder: (context, ref, child) {
  final pages = _getRoutes(_path, ref.watch(authVM));
  return Navigator(
  ...
}
```

> **What is WidgetRef?**
>
> As we have seen, we can watch a provider's value by using a `ref` object of the `WidgetRef` type. The Riverpod documentation defines `WidgetRef` as an object that allows widgets to interact with providers. `WidgetRef` lets us access any provider inside our app.

We now understand the basics of providers and we know how to create and read a provider. In the next section, we will implement Riverpod as the state management solution in our application.

Implementing Riverpod in our app

Now that we know the basics of Riverpod, we will implement it in our application. Remember that we have already used it in our application while implementing the MVVM pattern; however, this time, we will make each concept clearer and follow a process so that you will be able to implement it in your new projects. We will start by installing a dependency. Although we can use the Riverpod package directly and build our own widgets, as we saw previously, the `flutter_Riverpod` package already provides useful widget bindings, meaning we will install `flutter_Riverpod` as outlined in the preceding section. Open the `chapter8_start/pubspec.yaml` file and add the following under dependencies:

```
flutter_Riverpod: ^1.0.3
```

Then, run the following command:

```
flutter pub get
```

Note that if you are using `flutter_hooks`, you should install `hooks_Riverpod` instead. This includes extra features related to integration with Hooks.

Now, we are ready to start. Open `chapter8_start/lib/main.dart` and wrap the root widget with the `ProviderScope` widget as follows:

```
void main() async {
  WidgetsFlutterBinding.ensureInitialized();
  await Hive.initFlutter();
  runApp(ProviderScope(child: MyApp()));
}
```

`ProviderScope` is the widget that stores the state of all the providers we create in our application.

Next, we have our view models. All our view models use Riverpod to share data with a view. So, let's look at our `AuthVM`. Open `chapter8_start/lib/app/view_models/auth.vm.dart`. Our `AuthVM` is a change notifier and if we look at the bottom, we see the following:

```
final authVM = ChangeNotifierProvider<AuthVM>((ref) =>
AuthVM());
```

As we learned previously, here, we are creating a provider, and this time, it's ChangeNotifierProvider, as our AuthVM is a change notifier. We can see here that we have to access ref (a WidgetRef object) if we want to access other providers. Once the provider is created, it's time to use it. First, in the login view, when the login button is pressed, we want to call the login function of AuthVM. Let's see how we go about that. Open chapter8_final/lib/app/views/login.view.dart and look at line 37:

```
Consumer(builder: (context, ref, child) {
    ...
})
```

The consumer widget that comes from the Riverpod package provides a builder callback that has three arguments. The first is the widget's BuildContext. The second is the WidgetRef object, which we will use next to read and watch our provider. Finally, a widget, any widget passed to the child parameter of the Consumer widget, is available here as the third argument. We can use this to show other widgets that don't need to be rebuilt during the consumer builder rebuild. Next, we will look at the onPressed callback of the login's ElevatedButton widget:

```
onPressed: () {
  if (ref
      .read(authVM)
      .login(email: _email.text, password: _password.text)) {
    //logged in
  } else {
    // error
    debugPrint(ref.read(authVM).error);
  }
},
```

Here, we are using the ref object to access our authentication view model. We are also using ref.read() instead of listen or watch because this happens when we click a button where we don't want to listen to or observe changes. This does not happen during building.

Next, when the app is first running, to decide the routes where we want to check whether the user is already logged in, we can listen to the AuthVM provider. Open chapter8_start/lib/app/routes/router_delegate.router.dart and, inside the build method, we have the following:

```
  return Consumer(builder: (context, ref, child) {
```

```
final pages = _getRoutes(_path, ref.watch(authVM));
return Navigator(
  ...
}
```

Again, we are using the same `Consumer` widget that we talked about previously. This time, however, we are using `ref.watch(authVM)` to watch `authVM` instead of reading, as here, whenever the state changes, we want the navigator to rebuild and display the routes accordingly. We are passing `authVM` to the `_getRoutes` function, where, if you look, based on whether or not the user is logged in, we display different routes.

Let's look at the course list view model. Open `chapter8_start/lib/app/view_models/course_list.vm.dart`. Here, we define `StateNotifier` that stores the list of courses, right now loaded from a JSON asset, but later, when we integrate with external services such as Appwrite or Firebase, we will get courses from the database. For the state notifier, we define the provider using `StateNotifierProvider`, as follows:

```
final courseListVM =
  StateNotifierProvider<CourseListVM, List<CourseVM>>(
    (ref) => CourseListVM()
  );
```

`StateNotifierProvider` requires two generic types: first, the `StateNotifier`, and then the type of data that `StateNotifier` stores. Let's now build a course list widget that displays the list of courses. Create a `chapter8_start/lib/app/widgets/course_list.widget.dart` file. We will start by creating a simple stateless widget:

```
import 'package:flutter/material.dart';
class CourseList extends ConsumerWidget {
  const CourseList({Key? key}) : super(key: key);
  @override
  Widget build(BuildContext context, WidgetRef ref) {
    ...
  }
}
```

As you can see, we are subclassing our `CourseList` widget from `ConsumerWidget` so that we can get the `WidgetRef` object in our `build` function. This is another way in which we can get `WidgetRef` and access the providers. Here, we will watch `courseListVM` as follows:

```
Widget build(BuildContext context, WidgetRef ref) {
  final courses = ref.watch(courseListVM);
  …
}
```

Using the courses we get from `courseListVM`, we can build our `GridView` to display the courses as follows. Return the following `GridView` from the `build` method:

```
return GridView.builder(
  gridDelegate: SliverGridDelegateWithFixedCrossAxisCount(
    crossAxisCount: 2
    itemCount: courses.length,
    itemBuilder: (BuildContext context, int index) {
      final course = courses[index];
      return CourseCard(
          id: course.course.id,
          image: course.image,
          title: course.title,
          onActionPressed: () {},
          description: course.description);
    },
);
```

Finally, `crossAxisCount` is set to 2, although we can update this to make it a bit more responsive to the width as follows. First, get the width from `MediaQuery` inside the `build` method:

```
final width = MediaQuery.of(context).size.width;
```

Next, update `crossAxisCount` as follows:

```
crossAxisCount: width > ScreenSizes.xl
    ? 4
    : width > ScreenSizes.md
        ? 2
        : 1
```

So, for any screen size larger than 1,280 px, we want to use four columns; above 768 px, we want to use two columns; and for anything less, we will use a single column.

Summary

In this chapter, we learned the basics of state management. We started by learning about what a state is. Then, we learned about why we may need a state management solution in our application and what options are available in Flutter to manage the application state. We briefly looked at the various state management solutions available. Then, we introduced Riverpod, one of these state management solutions. We learned about the basics of Provider, a concept heavily used by Riverpod to build a state management solution. Finally, we implemented Riverpod in our application. In the next chapter, we will learn all about Appwrite and integrate it into our application.

9

Integrating Appwrite

In the previous chapter, we learned about state management and what tools and techniques are available to manage the state of an application. We looked at various packages available for state management and implemented Riverpod in our application.

In this chapter, we are going to cover integration with Appwrite. We will learn what Appwrite is and why it is useful for building Flutter applications. We will learn how to install Appwrite – then, we will set up an Appwrite project and install and configure Appwrite's SDK. Finally, we will integrate Appwrite authentication and its database in our application.

By the end of this chapter, you will be able to understand Appwrite and integrate it with any of your applications.

In this chapter, we will cover the following topics:

- What is Appwrite?
- Installing Appwrite
- Setting up an Appwrite project
- Installing and configuring Appwrite's SDK for Flutter
- Authenticating with Appwrite
- Using Appwrite's database to persist data

Technical requirements

The technical requirements for this chapter are as follows:

- Flutter version 3.0 or later installed and running
- Visual Studio Code or Android Studio
- Google Chrome browser
- Local or virtual machine with at least 2 GB free memory with Docker installed

You can download the code samples for this chapter from the book's official GitHub repository at `https://github.com/PacktPublishing/Taking-Flutter-to-the-Web`. The starter code for this chapter can be found inside the `chapter9_start` folder.

What is Appwrite?

Appwrite is an awesome open source backend as a service for web and mobile applications, including Flutter. It provides a wide range of features that are essential for building modern applications. Modern applications are content-rich, collaborative, and sync across devices. All of this would not be possible without a solid backend. Appwrite is an example of this kind of solid backend. Also, I think Appwrite is a great complement to Flutter, as they share some of the same qualities. First, both are open source; second, both focus on making the developer experience awesome, and finally, both are simple to work with and integrate well together. Appwrite provides various services, such as authentication, databases, storage, and cloud functions, all of which are accessible via a REST API or a real-time API using WebSocket. Don't worry – we don't have to work from scratch, as Appwrite already provides awesome SDKs for many platforms and languages, including Flutter.

As Appwrite is a robust backend solution maintained by a solid organization and a great open source community, it allows us to focus on our application by taking care of all the repetitive tasks required for developing a modern application.

Now that we know what Appwrite is and why it is an important tool for our application, in the next section, we will learn how to install it.

Installing Appwrite

Appwrite doesn't yet have a cloud solution, so to use it, we will have to host it ourselves. That also means we can easily install and run it locally to see it in action and use it during development. There are a couple of ways we can install Appwrite. We will look at each of them in this section. At the time of writing, the latest version of Appwrite is version 1.0.1, and it can run on any machine that supports Docker. The only requirement to install Appwrite is to install Docker and Docker Compose.

Installing Appwrite in DigitalOcean servers

This is the quickest and easiest method to install Appwrite. With this one-click setup, you can quickly spin up an Appwrite server that is accessible over the internet. For this, you will need a DigitalOcean account (`https://digitalocean.com`). Once you have the account, visit `https://marketplace.digitalocean.com/apps/appwrite` and click on **Create Appwrite Droplet**:

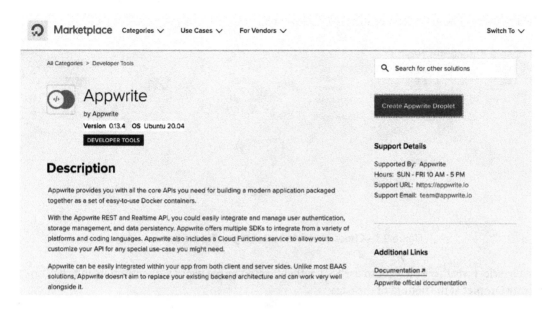

Figure 9.1 – The DigitalOcean marketplace app for Appwrite

You will be redirected to a page where you can select a specification for the server. I recommend choosing at least 2 GB of RAM:

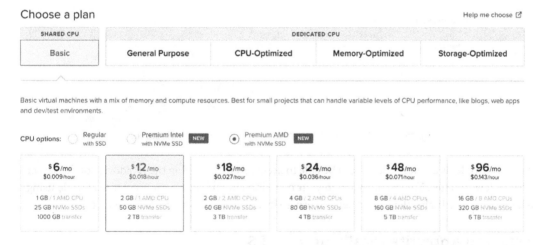

Figure 9.2 – Choosing a plan for the server

Then, you can choose the data center region for your server:

Choose a datacenter region

Figure 9.3 – Choosing a data center region for the server

Finally, select whether you want to use SSH authentication or password authentication and click on **Create Droplet** at the bottom of the page:

Figure 9.4 – Choosing the authentication method for the server

Once the droplet is created, you should be able to access Appwrite using the IP address of the server within a few minutes. You can also assign your own domain and access the Appwrite console using that domain. This is also the recommended way to install and run Appwrite in production.

Installing Appwrite locally or in a VPS

To install Appwrite, we need to install Docker first. Follow the official guide, available from `https://docs.docker.com/get-docker/`, to install Docker on the device where you are planning to install Appwrite. Then, we can use a single command to install Appwrite.

If you are on Linux or macOS, run the following command:

```
docker run -it --rm \
    --volume /var/run/docker.sock:/var/run/docker.sock \
    --volume "$(pwd)"/appwrite:/usr/src/code/appwrite:rw \
    --entrypoint="install" \
    appwrite/appwrite:1.0.1
```

If you are, however, on Windows, use one of the following commands based on whether you are using PowerShell or Command Prompt.

Use the following for PowerShell:

```
docker run -it --rm ,
    --volume /var/run/docker.sock:/var/run/docker.sock ,
    --volume ${pwd}/appwrite:/usr/src/code/appwrite:rw ,
    --entrypoint="install" ,
    appwrite/appwrite:1.0.1
```

Use the following for Command Prompt:

```
docker run -it --rm ^
    --volume //var/run/docker.sock:/var/run/docker.sock ^
    --volume "%cd%"/appwrite:/usr/src/code/appwrite:rw ^
    --entrypoint="install" ^
    appwrite/appwrite:1.0.1
```

This command should download and install Appwrite and it should be accessible over localhost. If you are installing on a **virtual private server** (**VPS**), you should be able to access it via the IP address or domain that you have assigned.

Well, now we know how to install Appwrite. If you want to learn more or are confused by any of the steps, you can also visit the official installation documentation at `https://appwrite.io/docs/installation`. In the next section, we will learn how to install and configure Appwrite's SDK for Flutter.

Setting up an Appwrite project

To be able to integrate Appwrite with our application, we need to set up an Appwrite project in the Appwrite server first. Follow these steps to set up an Appwrite project:

1. When you install Appwrite and access it, you will first be redirected to the **Sign Up** page. Enter the details and click **Sign Up**:

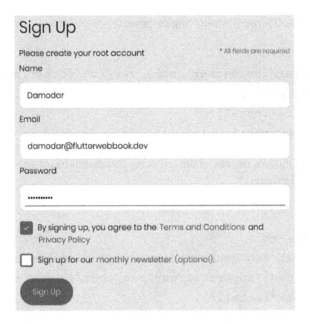

Figure 9.5 – Signing up for the Appwrite console

This should now create a new account and redirect you to the console:

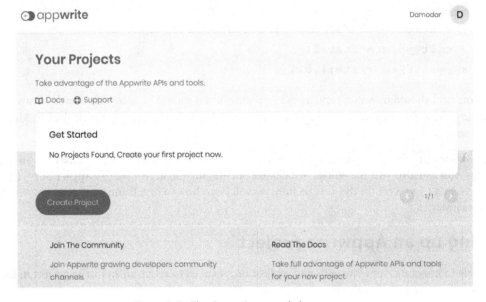

Figure 9.6 – The Appwrite console home page

2. Click on the **Create Project** button, then on the dialog box that appears – enter a name for your project and click on the **Create** button:

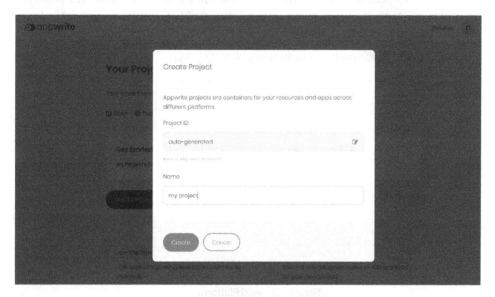

Figure 9.7 – The Create Project dialog

You can also switch the **Project ID** field from the autogenerated ID to a custom ID by clicking on the edit icon. You should now be redirected to the project home page:

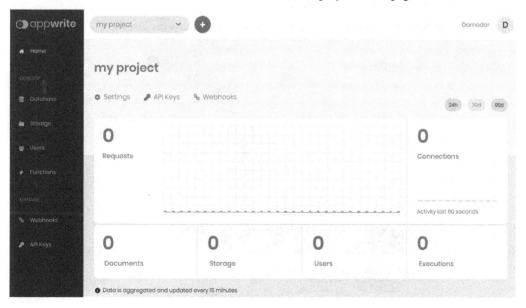

Figure 9.8 – The project home page

3. Scroll down to find the **Add Platform** button.

    This is where we need to add platforms so that our Flutter and other client-side projects can connect to our Appwrite server. As we are developing for Flutter for Web, we should add a web platform instead of a Flutter platform.

4. Click on the **Add Platform** button and select **New Web App**:

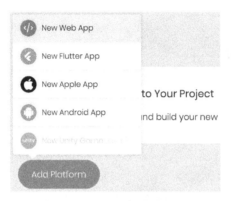

Figure 9.9 – Add Platform

5. In the dialog box that appears, provide a recognizable name for your platform, and for the hostname, add the domain that your app will run in. Right now, we are testing it locally, so the hostname to use would be `localhost`.

6. Click the **Register** button to add the new platform to the platforms list:

Figure 9.10 – The New Web App platform dialog

Now that we have added a platform, all we need now is the project ID and the endpoint, which we will use later to configure our SDK in our applications.

7.   To get these details, click on the **Settings** link from the project home page at the top or the bottom left on the sidebar:

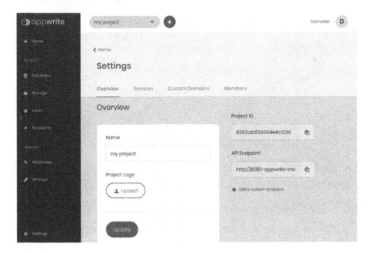

Figure 9.11 – Project settings

8.   Note down the **Project ID** and **API Endpoint** values.

Now that we have this information, we will install and configure Appwrite's SDK for Flutter in our project in the next section.

Installing and configuring Appwrite's SDK for Flutter

Integrating Appwrite in Flutter applications is very simple using the official Flutter SDK. So, in this section, we are going to install and configure Appwrite's Flutter SDK in our application by following these steps:

1.   First, let's add the Appwrite package under `dependencies` in the `pubspec.yml` file:

```
dependencies:
        appwrite: ^8.0.0
```

Remember to run `flutter pub get` if it's not automatically run by the IDE to get the dependencies.

2.   Next, we will import and configure Appwrite's SDK. We will do that by creating the `lib/infrastructure/res/appwrite.service.dart` file:

```
import 'package:appwrite/appwrite.dart';
```

3. Configure the SDK by providing the endpoint and project ID. You should be able to get both of these values for your Appwrite installation from your console, as we saw in the previous section:

```
class AppwriteService {
  static AppwriteService instance = appwriteService;
  final Client client;
  AppwriteService() : client = Client() {
   client.setEndpoint('https://demo.appwrite.io/v1')
     .setProject('projectId');
  }
}
final appwriteService = AppwriteService();
```

Now that we have installed and configured our SDK, we can move on to using the services provided by Appwrite. In the next section, we will start by implementing the authentication service provided by Appwrite to authenticate our users.

Authenticating with Appwrite

Appwrite provides an authentication service that we can use to authenticate our application users. Appwrite's authentication supports anonymous authentication, email- and password-based authentication, as well as OAuth authentication using lots of different OAuth providers, including Google, Facebook, and LinkedIn. We will start by implementing anonymous login, which is a very useful feature if you don't want to keep track of any user details. We are already using the MVVM pattern, so we can implement anonymous login in our auth view model. Follow these steps:

1. Open lib/app/view_models/auth.vm.dart and first import and instantiate the account service provided by Appwrite.

2. Import package:appwrite/appwrite.dart:

```
class AuthVM extends ChangeNotifier {
  bool isLoggedIn = false;
  String error = '';
  UserVM? User;
  final Account account = Account(AppwriteService.
instance.client);

  ...
}
```

3. Next, add an anonymous login method:

```
Future<bool> anonymousLogin() async {
  if (isLoggedIn) {
    error = 'Already logged in';
    return false;
  }
}
```

4. Finally, to log the user in anonymously, call `account.createAnonymousSession()`. Let's do that with proper error handling:

```
error = '';
try {
  await account.createAnonymousSession();
  final apUser = await account.get();
  user = UserVM(id: apUser.$id, email: 'N/A', name:
'Anonymous User');
  isLoggedIn = true;
  notifyListeners();
  return true;
} on AppwriteException catch (e) {
  error = e.message ?? 'Unknown error';
  return false;
}
```

Now that we have a function to log in anonymously, let's use that in our login view by following these steps:

1. Open `lib/app/views/login.view.dart` and copy and update the existing login button, as follows:

```
Consumer(builder: (context, ref, child) {
  return ElevatedButton(
    onPressed: () async {
      if (await ref
          .read(authVM)
          .anonymousLogin()) {
        //logged in
      } else {
        // error
        debugPrint(ref.read(authVM).error);
```

```
        }
      },
      child: Text("Anonymous Login"),
    );
  }),
```

If we run the application now and tap the **Anonymous Login** button on the login page, we should be able to log in to the application without providing an email or password.

2. Next, let's update the `login` method to use Appwrite's authentication instead. Open `lib/app/view_models/auth.vm.dart` and update the `login` method as follows:

```
Future<bool> login({required String email, required
String password}) async {
  try {
    await account.createEmailSession(email: email,
        password: password);
    final apUser = await account.get();
    user = UserVM(id: apUser.$id, email: apUser.email,
        name: apUser.name);
    isLoggedIn = true;
    notifyListeners();
    return true;
  } on AppwriteException catch (e) {
    error = e.message ?? e.toString();
    notifyListeners();
    return false;
  }
}
```

3. To log in, we use the `account.createEmailSession` function of Appwrite's account service and pass the email and password. If successful, we will get back a session object, which we are not required to store. If it fails, we get `AppwriteException`, which we use to catch and show the error.

    Once `createEmailSession` is successful, we call `account.get()` to get the account details and update `UserVM` accordingly. In Appwrite, creating a session is the way to provide a valid session for the user in the Appwrite server. The user is authenticated via cookies, and cookies are handled automatically by the SDK. Once the user logs in, all requests made to Appwrite from the SDK will automatically pass the valid cookies, which the server uses to validate the user.

Finally, let's see how we register new users in Appwrite. In the auth view model, we already have a function called `register`. Let's update it as follows:

```
Future<bool> register({required String name, required String
email, required String password}) async {
  try {
    await account.create(userId: 'unique()', name: name,
        email: email, password: password);
    await login(email: email, password: password);
    return true;
  } on AppwriteException catch (e) {
    error = e.message ?? e.toString();
    notifyListeners();
    return false;
  }
}
```

In the preceding code block, we first called `account.create()`, which creates a new user in Appwrite. However, in Appwrite, creating a user doesn't automatically create a session for the user. So, to create a valid session, we again called the `login()` method. In `account.create()`, we passed `userId: 'unique()'`. This tells Appwrite to automatically generate a unique ID for our user. We could, however, provide our own custom ID as well.

That is all for now on the topic of authenticating with Appwrite. To learn more about Appwrite's account service, you can head over to `https://appwrite.io/docs/client/account`. In the next section, we will learn about Appwrite's database service.

Using Appwrite's database to persist data

Apart from authentication, another popular and common service required for applications is a database to store and share data. Appwrite provides a highly customizable database with a flexible permission system. In this section, we will learn how to use Appwrite's database to build our Flutter Academy application. We will start by creating a collection where we will add courses and display those courses in our application. Let's begin by setting up the course database.

Creating a Database and a collection

Appwrite's database provides an interface similar to a over the relational database. We create collections and documents, where collections provide a structured interface to save data. So, we will start by creating a collection for our courses.

Collections can be set up either from Appwrite's console or via Appwrite's server-side SDKs. If you want to do it from the console, it's simple to follow the GUI. GUIs tend to change more often than the API and SDK itself – that is why here, we will write a simple Dart script using Appwrite's server-side SDK to set up our collection and insert seed data. To set up the server-side SDK, we need to get an API key. To get that, do the following:

1. Go to the Appwrite console and click on **API Keys** on the left sidebar.

2. Once the **API Keys** panel opens, click on the **Add Key** button.

3. Give your key an appropriate name and then select the **databases.write**, **databases.read**, **collections.write**, **collections.read**, **documents.write**, **documents.read**, **attributes.write**, **attributes.read**, **indexes.write**, and **indexes.read** scopes. Click on the **Create** button:

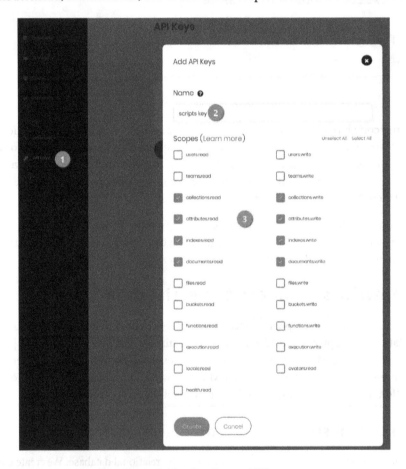

Figure 9.12 – Creating an API key

4. Click on **Show Secret** and copy the API key:

Figure 9.13 – Showing and copying the secret

5. Next, open `chapter9_start/scripts/config.dart` and update the API key constant:

```
const String API_KEY =
'7cbf4c353e5ac0fb674909084ed...479f2eed5c8a49ee';
```

6. Then, open `chapter9_start/scripts/pubspec.yaml` and add Appwrite's Dart SDK:

```
dependencies:
  dart_appwrite: 7.0.0
```

7. Then, run the following command to get the dependencies:

```
flutter pub get
```

8. Then, open `scripts/main.dart` and import and initialize Appwrite's Dart SDK:

```
import 'package:dart_appwrite/dart_appwrite.dart';
import 'config.dart' as config;

final client = Client()
    .setEndpoint('https://[YOUR_APPWRITE_DOMAIN]/v1/')
```

```
                        .setProject('flutter-academy')
                        .setKey(config.API_KEY);
```

Make sure to provide a proper endpoint for your installation of Appwrite.

Next, let's add a function that will create our database and then collection and attributes.

Adding a function to create Database

First, we need to create a database to store our collections. Databases are like a group of collections. Let's do it as follows:

```
Future<void> createDatabase() async {
  final db = Databases(client);
  await db.create(databaseId: 'flutter_academy_db', name:
'Flutter Academy DB');
}
```

Adding a function to create collections and attributes

Next we need to create a collection as follows:

```
Future<void> createCollection() async {
  final db = Database(client);
await db.createCollection(
    databaseId: 'flutter_academy_db',
    collectionId: 'courses',
    name: 'Courses',
    documentSecurity: true,
    permissions: [
      Permission.create(Role.users()),
    ],
  );
}
```

Notice that documentSecurity is set to true. Doing this will allow us to specify both the collection and document permissions and users with permission, in one of the level, can gain access to the resources. We also specify the collection permissions, where we only have set create permission. With Role.users(), anyone having a valid session in the server can create new document in the courses collection. To learn more about Appwrite's permissions, visit https://appwrite.io/docs/permissions.

Now that we have added the collection, we need to add individual attributes. We will add the title, description, image, and status as a string, and `published_date` as the integer attributes. Add the following lines of code to create all these attributes:

```
await db.createStringAttribute(
     databaseId: 'flutter_academy_db',
     collectionId: 'courses',
     key: 'title',
     size: 255,
     xrequired: true);
  await db.createStringAttribute(
     databaseId: 'flutter_academy_db',
     collectionId: 'courses',
     key: 'description',
     size: 1000,
     xrequired: false);
  await db.createStringAttribute(
     databaseId: 'flutter_academy_db',
     collectionId: 'courses',
     key: 'image',
     size: 255,
     xrequired: false);
  await db.createStringAttribute(
     databaseId: 'flutter_academy_db',
     collectionId: 'courses',
     key: 'status',
     size: 20,
     xrequired: false,
     xdefault: 'Draft');
  await db.createIntegerAttribute(
     databaseId: 'flutter_academy_db',
     collectionId: 'courses',
     key: 'published_date',
     xrequired: true);
```

Finally, we need to add an index for the status, as we would want to query by status and, for example, maybe only show published courses:

```
await Future.delayed(Duration(seconds: 5));
await db.createIndex(
    databaseId: 'flutter_academy_db',
    collectionId: 'courses',
    key: 'status_index',
    type: 'key',
    attributes: ['status']);
```

We are delaying by 5 seconds to make sure that the collection attributes are successfully created, as Appwrite creates attributes in the background to make the API responsive.

Next, we will add some course documents from our script so that we have something to display in our application.

Adding course documents and listing them on the app

This is usually a part of the real-world version of this project where we would create a separate admin panel where admins or authors can log in and create and manage courses. However, that's beyond the scope of this book.

Follow these steps to add course documents:

1. Start by making a list of courses that we can add to the database. Prepare a list in the following format:

```
final courses = [
    {
        "title": "Flutter Beginners",
        "description":
            "Awesome course for Flutter beginners to learn
                the basics of Flutter framework",
        "image": "https://image-placeholder.com/
            image.png",
        "status": "published",
        "published_date": DateTime.now().
            millisecondsSinceEpoch,
    },
    ...
];
```

2. Add more courses with the same keys as the attributes we have in the database. Make sure to make some course drafts so that we can learn about how to only show published courses later in this section.

3. Next, add a function named `addCourses` to the same file. Use Appwrite's database service again to insert those courses into the course collection that we previously created:

```
Future<void> addCourses() async {
  final db = Database(client);
  for (final course in courses) {
      await db.createDocument(
      databaseId: 'flutter_academy_db',
      collectionId: 'courses',
      documentId: 'unique()',
      data: course,
    );
  }
}
```

4. We now need to call methods to create a collection and add courses in the `main` function:

```
void main(List<String> arguments) async {
  await createDatabase();
  await createCollection();
  await addCourses();
}
```

5. To run this script from Command Prompt or Terminal, while in the chapter's directory, run the following command:

`dart scripts/main.dart`

If the script completes without any errors, then we should have a course collection and a few courses added to it. When you log in to your console, you should be able to view your new collection and documents there:

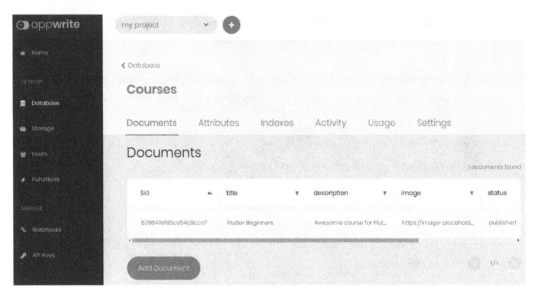

Figure 9.14 – The Courses collection and its documents

Next, we will now display these courses in our application, instead of using a local JSON file to load the data.

Open /lib/infrastructure/res/course.service.dart and first, to use Appwrite's database, we need to instantiate its database service. Let's do that in our CourseService class by following these steps:

1. First, import the Appwrite package:

    ```
    import 'package:appwrite/appwrite.dart';
    ```

2. Then, instantiate the database service:

    ```
    class CourseService {
      final database = Database(AppwriteService.instance.
    client);

        ...
    }
    ```

3. Then, update the getCourses function to use Appwrite's database to list the courses, as follows:

    ```
    Future<List<Course>> getCourses() async {
      final docs = await database.listDocuments(
          databaseId: 'flutter_academy_db',
          collectionId: 'courses');
    ```

```
      return docs
          .convertTo((data) => Course.fromMap(Map<String,
  dynamic>.from(data)));
  }
```

Here, we are simply calling the `listDocuments` method of the database service and passing our collection ID. In the next line, we are converting the data we get to a `Course` object.

Next, if we only want to get the published courses, we can do that by passing a query.

4. Let's update the `listDocuments` method to use a query to only load the published courses:

```
final docs = await database.listDocuments(
  databaseId: 'flutter_academy_db',
  collectionId: 'courses',
  queries: [
    Query.equal('status', 'published'),
  ]
);
```

Queries are simply a well-formatted list of strings; however, we don't have to understand or know the format. Appwrite's SDKs provide a `Query` helper class, which we can use to build our queries as we did in the preceding code. This now only loads the courses that have a status of `published`. The good thing about using MVVM is that although we have updated the course service to load data from Appwrite and the `Course` model has changed, it doesn't affect the UI. We don't have to change anything in the UI.

Next, we will update our watchlist functionality so that we can save the user's watchlist in Appwrite instead of saving it locally.

Updating the watchlist functionality using Appwrite's database

Here, again, we will first start by creating a new collection to hold the user's watchlist. We will use the same script as we set previously. Let's get started:

1. Add a new function called `createWatchlistCollection()` in `scripts/main.dart`:

```
void createWatchlistCollection() async {
  final db = Databases(client);
  await db.createCollection(
    databaseId: 'flutter_academy_db',
    collectionId: 'watchlist',
    name: "Watchlist",
```

```
      documentSecurity: true,
      permissions: [
        Permission.create(Role.users()),
      ],
    );
    await db.createStringAttribute(
        databaseId: 'flutter_academy_db',
        collectionId: 'watchlist',
        key: 'userId', size: '36', xrequired: true);
    await db.createStringAttribute(
        databaseId: 'flutter_academy_db',
      collectionId: 'watchlist', key: 'courseId',
      size: '36', xrequired: true);
}
```

As we will be querying this collection to filter by courseId, we also need to create an index for it.

2. Add the following line after creating the courseId attribute:

```
await Future.delayed(Duration(seconds: 5))
await db.createIndex(
  databaseId: 'flutter_academy_db',
  collectionId: 'watchlist',
  key: 'course_id_index',
  type: 'key',
  attributes: ['courseId']);
```

3. In the main function, we remove the calls to previous methods and call this instead:

```
void main(List<String> arguments) async {
  await createWatchlistCollection();
}
```

4. Finally, run the main.dart script again from your terminal as you did previously:

dart scripts/main.dart

You should be able to see your new collection in the **Database** section of the Appwrite console. The attributes and indexes should be available:

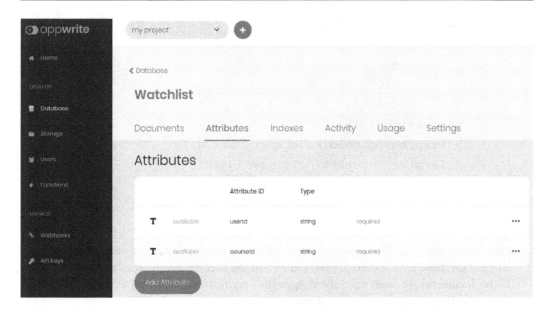

Figure 9.15 – The Watchlist collection and its Attributes

Next, we will update the watchlist service to use Appwrite by following these steps:

1. Open `lib/infrastructure/res/watchlist.service.dart` and import the Appwrite package first, as we did previously. Then, update the class with new properties for the database service and collection ID, as follows:

```
import 'package:appwrite/appwrite.dart';
class WatchlistService {
  final databaseId = 'flutter_academy_db';
  final collectionId = 'watchlist';
  final db = Database(AppwriteService.
      instance.client);
  ...
}
```

2. Next, update the `addToWatchlist` function as follows to use Appwrite's database service:

```
Future<void> addToWatchlist(String id, String userId)
async {
  db.createDocument(
    databaseId: databaseId,
    collectionId: collectionId,
```

```
      documentId: 'unique()',
      data: {
        'userId': userId,
        'courseId': id,
      },
      permissions: [
        Permission.read(Role.user(userId)),
        Permission.write(Role.user(userId))
      ],
    );
  }
```

Here, make sure to add the proper permissions for the user so that when we list documents, we only get those for the current user, and so that the user has proper read and write access to the documents. To learn more about Appwrite's permissions, head over to the permissions docs at https://appwrite.io/docs/permissions.

3. Next, to remove a course from the watchlist, we first query the watchlist collection by courseId, then delete the record from the watchlist. We don't need to add userId to the query, as that's already filtered by Appwrite's permissions system:

```
Future<void> removeFromWatchlist(String id) async {
  final doc = await db.listDocuments(
      databaseId: databaseId,
      collectionId: collectionId, queries:
          [Query.equal('courseId', id)]);
  if (doc.total > 0) {
    await db.deleteDocument(
        databaseId: databaseId,
        collectionId: collectionId, documentId:
            doc.documents[0].$id);
  }
}
```

4. Finally, to get the courses in the watchlist, we first get the course IDs from the watchlist collection and then query the course collection for matching IDs:

```
Future<List<Course>> getWatchlist() async {
  final docList = await db.listDocuments(databaseId:
databaseId, collectionId: collectionId);
```

```
final docIds = docList.documents.map((doc) =>
    doc.$id).toList();
final courseList = await db.listDocuments(
    databaseId: databaseId,
    collectionId: 'courses', queries:
        [Query.equal('\$id', docIds)]);
return courseList
    .convertTo((doc) => Course.fromMap(Map<String,
        dynamic>.from(doc)));
}
```

The convertTo function, available in Appwrite's DocumentList and Document models, allows us to convert the database documents we receive into the appropriate data model.

This concludes our database integration.

Besides authentication and the database, Appwrite has various other services, such as user management, cloud functions, and storage, which help you to build robust and rich applications. To learn more about the various Appwrite services, you can visit the official documentation at https://appwrite.io/docs.

Summary

In this chapter, we introduced an open-source backend as a service, Appwrite. We then looked at how Appwrite can help us to focus on building the important parts of an application and delegate repetitive tasks, such as creating a database API for remote data persistence and authentication. Then, we learned how to install Appwrite and use it with our application to authenticate our users, as well as load and save our application data persistently.

In the next chapter, we will learn all about another similar service, Firebase by Google, and integrate it into our application instead of Appwrite.

10
Firebase Integration

In the previous chapter, we learned about Appwrite, an open source backend-as-a-service integration with our application. We learned how to install and configure Appwrite and use it in our application to authenticate our users and save their wishlists.

In this chapter, we are going to learn about another popular backend-as-a-service solution for Flutter applications and that is Google's Firebase. We will learn what Firebase is and how we can use it with our Flutter web application.

By the end of this chapter, you will understand what Firebase is, how to integrate it with your Flutter application, and how to use its various features.

In this chapter, we will cover the following topics:

- What is Firebase?
- Getting started with Firebase
- Installing and configuring Firebase's SDKs for Flutter
- Authentication with Firebase
- Using Firebase Firestore to persist data

Technical requirements

The technical requirements for this chapter are as follows:

- Flutter version 3.0 or later installed and running
- Visual Studio Code or Android Studio
- Google Chrome browser
- Google account to access Firebase services
- Node.js and npm to install and use Firebase tools

You can download the code samples for this chapter and other chapters from the book's official GitHub repository at `https://github.com/PacktPublishing/Taking-Flutter-to-the-Web`. The starter code for this chapter can be found inside the `chapter10_start` folder.

What is Firebase?

Firebase is a backend as a service for web and mobile applications. Firebase brings a great set of features and tools for web and mobile application developers. Firebase is a proprietary service offered by Google. It provides various services, such as Authentication, Realtime Database, Storage, Cloud Functions, Analytics, and Crashlytics. All of these services are accessible via their various SDKs.

Firebase provides official SDKs for Flutter and integrates well with Flutter. Firebase can be a great choice for developers when they want a robust backend but don't want to spend too much time and cost developing it. Though Firebase has a free tier, the cost of Firebase increases as your application's use rises. Make sure to go through the pricing (`https://firebase.google.com/pricing`) to understand more about how much each of the services of Firebase costs.

As Firebase provides various essential services for developers, and both Firebase and Flutter are Google products, they can be a great fit together. Now that we know what Firebase is and how it's beneficial for Flutter developers, in the next section, we will learn how to get started with Firebase.

Getting started with Firebase

Getting started with Firebase is quite simple. All you need is a Google email address. It can be your free Gmail or your Google Workspace account. First, visit `https://console.firebase.google.com` and make sure to log in with your Google email address.

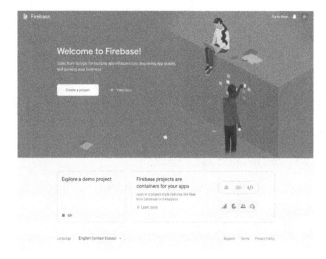

Figure 10.1 – The Firebase console

Once you log in, you will see the preceding console. If you have worked with Firebase before, you might have projects there already. We will start by creating a new project for our application. Click on the **Create a project** button. You will be directed to a page where you can provide a name for your project and optionally a custom ID, as shown in the following screenshot:

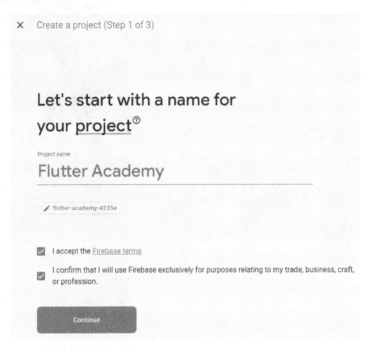

Figure 10.2 – New project – project name

On the next screen, you can choose to enable or disable analytics and click **Continue**. Once the project is created, you will be redirected to the project's overview page, as shown in the following screenshot:

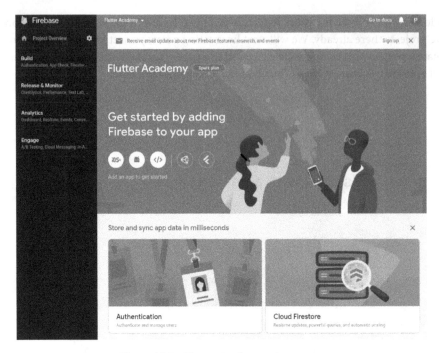

Figure 10.3 – The project's overview page

On the left-hand side, you can see a sidebar with various menu items grouped into different categories. There are tons of services that Firebase provides. You can use any and all of these services while building your application. Some services, such as Cloud Functions, will not be accessible to you until you activate billing in your account. You will not have to pay yet since there is a free tier, but to access Cloud Functions, you will still need to activate billing.

Though there is a wide variety of services that will be useful to you while developing mobile or web applications, we will mostly be using the Authentication and Firestore Database services in this book. The other services are beyond of the scope of this book. To learn more about other services and their documentation, head over to the official documentation site for Firebase at `https://firebase.google.com/docs`.

Now that we have our Firebase project, we are ready to move on to the next section, where we will integrate and configure our Flutter Academy application to use our newly created Firebase project.

Installing and configuring Firebase SDKs for Flutter

Though there are different ways we can configure a Flutter application for Firebase, we will use the easiest and most recommended method, which is using the FlutterFire CLI application. The application is written especially for Flutter projects. Let's get started:

1. First, install `flutterfire_cli` using the following command:

    `dart pub global activate flutterfire_cli`

    Doing this will install `flutterfire_cli` globally and the `flutterfire` command
    should be available in your command line. Confirm this by running the following command:

    `flutterfire -version`

    You should see the following output:

    `0.2.2+2`

2. `flutterfire_cli` uses the `firebase-tools` package internally, so we need to install
    that as well using the following command:

    `npm install -g firebase-tools`

    Doing this will install `firebase-tools` globally and the `firebase` command should be
    available in your command line. Confirm this by running the following command:

    `firebase -version`

    You should see the following output:

    `11.1.0`

3. Once the Firebase CLI is installed, run the following command and follow the instructions to
    log in to your Firebase account via the CLI:

    `firebase login`

    You will see instructions similar to the following. Follow and authenticate the CLI to use your
    account:

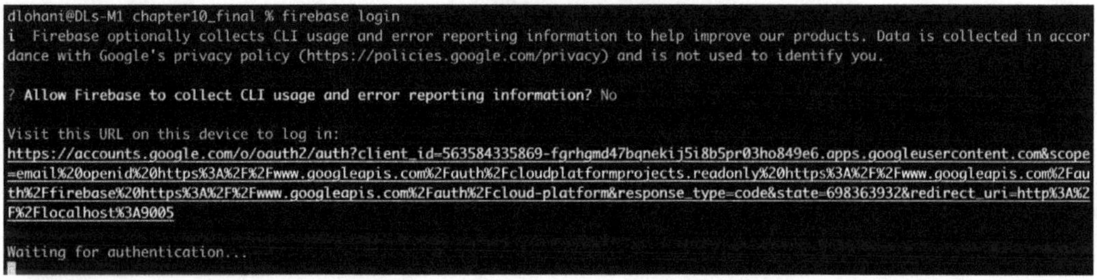

```
dlohani@DLs-M1 chapter10_final % firebase login
i  Firebase optionally collects CLI usage and error reporting information to help improve our products. Data is collected in accor
dance with Google's privacy policy (https://policies.google.com/privacy) and is not used to identify you.

? Allow Firebase to collect CLI usage and error reporting information? No

Visit this URL on this device to log in:
https://accounts.google.com/o/oauth2/auth?client_id=563584335869-fgrhgmd47bqnekij5i8b5pr03ho849e6.apps.googleusercontent.com&scope
=email%20openid%20https%3A%2F%2Fwww.googleapis.com%2Fauth%2Fcloudplatformprojects.readonly%20https%3A%2F%2Fwww.googleapis.com%2Fau
th%2Ffirebase%20https%3A%2F%2Fwww.googleapis.com%2Fauth%2Fcloud-platform&response_type=code&state=698363932&redirect_uri=http%3A%2
F%2Flocalhost%3A9005

Waiting for authentication...
```

Figure 10.4 – The Firebase login

4. After that, from your command line, navigate to the folder where you have saved your Flutter
    Academy project and run the following command:

```
flutterfire configure –project=[id-of-the-project-you-
created]
```

5. Then, follow the instructions presented on the screen:

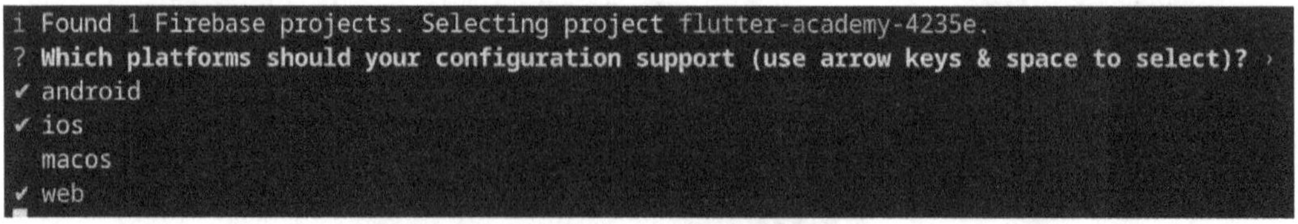

```
i Found 1 Firebase projects. Selecting project flutter-academy-4235e.
? Which platforms should your configuration support (use arrow keys & space to select)? ›
✔ android
✔ ios
  macos
✔ web
```

Figure 10.5 – Configuring the project

6. Once you select all the options, your project should be configured accordingly:

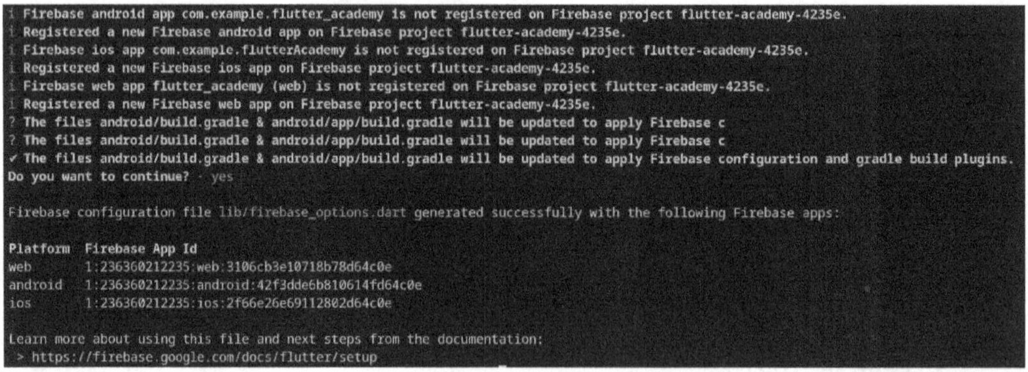

```
i Firebase android app com.example.flutter_academy is not registered on Firebase project flutter-academy-4235e.
i Registered a new Firebase android app on Firebase project flutter-academy-4235e.
i Firebase ios app com.example.flutterAcademy is not registered on Firebase project flutter-academy-4235e.
i Registered a new Firebase ios app on Firebase project flutter-academy-4235e.
i Firebase web app flutter_academy (web) is not registered on Firebase project flutter-academy-4235e.
i Registered a new Firebase web app on Firebase project flutter-academy-4235e.
? The files android/build.gradle & android/app/build.gradle will be updated to apply Firebase c
? The files android/build.gradle & android/app/build.gradle will be updated to apply Firebase c
✔ The files android/build.gradle & android/app/build.gradle will be updated to apply Firebase configuration and gradle build plugins.
Do you want to continue?  yes

Firebase configuration file lib/firebase_options.dart generated successfully with the following Firebase apps:

Platform  Firebase App Id
web       1:236360212235:web:3106cb3e10718b78d64c0e
android   1:236360212235:android:42f3dde6b810614fd64c0e
ios       1:236360212235:ios:2f66e26e69112802d64c0e

Learn more about using this file and next steps from the documentation:
 > https://firebase.google.com/docs/flutter/setup
```

Figure 10.6 – The configured project

7. Then, in the main.dart file, first, import firebase_core and firebase_options.
    dart, generated by flutterfire_cli:

    ```
    import 'package:firebase_core/firebase_core.dart';
    import 'firebase_options.dart';
    ```

8. Then, update the main function and add the Firebase initialization step with the provided
    Firebase options:

    ```
    await Firebase.initializeApp(
        options: DefaultFirebaseOptions.currentPlatform,
    );
    ```

9. Finally, we need to add the `firebase_core` package. Open `pubspec.yaml` and add the following under `dependencies`:

```
dependencies:
   firebase_core: ^1.17.1
```

That's it – we have configured our Flutter application to use our Firebase project. In the next section, we will look in detail into how to implement the Firebase Authentication service in our application.

Authentication with Firebase

Firebase provides an authentication service that app developers can use to authenticate their users. The Firebase Authentication service supports anonymous authentication, email and password-based authentication, and OAuth authentication with various OAuth providers, including Google, Facebook, and LinkedIn.

To implement authentication, we have to configure it in the Firebase console first. Follow these steps to configure the Firebase console:

1. In your console, click on **Build** | **Authentication** in the sidebar.

2. On the authentication page, click on the **Get started** button:

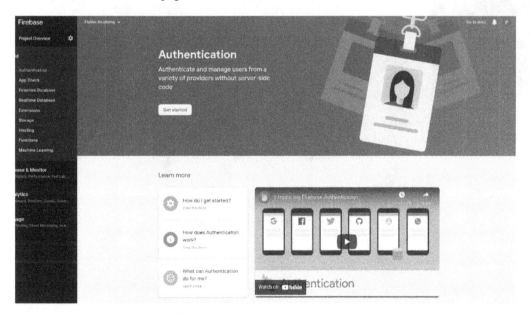

Figure 10.7 – Authentication – getting started

Initially, no providers are available.

3. Click on the **Email/Password** button, then enable the **Email/Password** authentication method, and click the **Save** button.

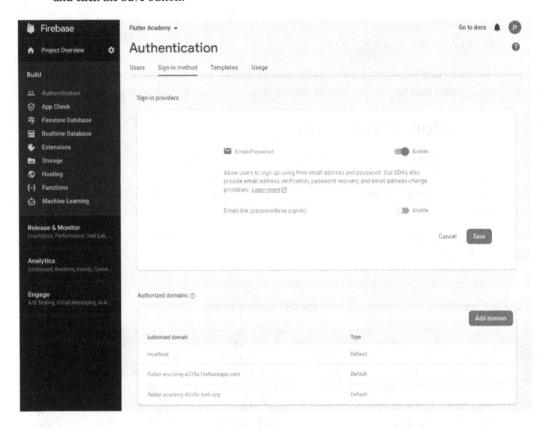

Figure 10.8 – Email and password-based authentication

4. Finally, click on the **Add new provider** button, and in the dialog that appears, select the **Anonymous** method:

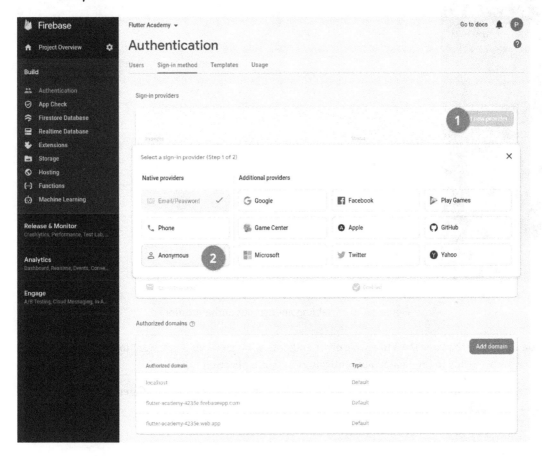

Figure 10.9 – Adding a new provider

5. Enable the **Anonymous** authentication method and click on the **Save** button:

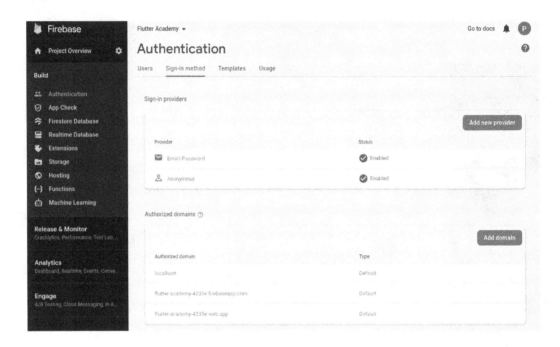

Figure 10.10 – Enabling anonymous authentication

You can follow the same steps to enable other authentication methods if you want to implement more. Now, we are ready to implement authentication in our application.

Let's start by adding the firebase_auth package in pubspec.yaml. Open the pubspec. yaml file and under dependencies, add firebase_auth, as follows:

```
dependencies:
  firebase_auth: ^3.3.19
```

We will then implement anonymous login, which is a very useful feature if you don't want to keep track of the details of your users. We are already using the MVVM pattern, so we implement anonymous login in our auth view model. Follow these steps:

1. Open lib/app/view_models/auth.vm.dart and import firebase_auth first:

    ```
    import 'package:firebase_aiuth/firebase_auth.dart';
    class AuthVM extends ChangeNotifier {
      final FirebaseAuth _auth = FirebaseAuth.instance;

      ...
    }
    ```

2. Next, add an anonymous login method:

```
Future<bool> anonymousLogin() async {
  if (isLoggedIn) {
    error = 'Already logged in';
    return false;
  }
}
```

3. Finally, to log a user in anonymously, we can call _auth.anonymousLogin(). Let's do that with proper error handling:

```
error = '';
await _auth.signInAnonymously();
```

Firebase provides a real-time update listener where we can listen to auth state changes. We implement that function to check whether or not the user is currently logged in. This will work with all the login methods.

4. In AuthVM, add a new method that listens to the auth state changes from Firebase Authentication, as follows:

```
Future<void> subscribe() async {
  _subscription =
  _auth.authStateChanges().listen((user) {
    if (user == null) {
      isLoggedIn = false;
      this.user = null;
      notifyListeners();
    } else {
      this.user = UserVM(
          email: user.email ?? 'N/A',
          name: user.displayName ?? 'N/A',
          id: user.uid);
      isLoggedIn = true;
      notifyListeners();
    }
  });
}
```

5. Also, add `StreamSubscription _subscription` as a property of `AuthVM` to be able to unsubscribe, as follows:

```
StreamSubscription<User?>? _subscription;
```

6. In the constructor of `AuthVM`, call this method so that the subscription is called when we start the application. Also, add a `dispose` method to unsubscribe or we will have a memory leak:

```
@override
void dispose() {
  _subscription?.cancel();
  super.dispose();
}
```

7. Remember to call the `subscribe` method from the constructor of `AuthVM`:

```
AuthVM() {
  subscribe();
}
```

Now that we have a function to log in anonymously, let's use that in our login view.

Open `lib/app/views/login.view.dart` and copy and update the existing login button, as follows:

```
Consumer(builder: (context, ref, child) {
  return ElevatedButton(
    onPressed: () async {
      if (await ref
          .read(authVM)
          .anonymousLogin()) {
        //logged in
      } else {
        // error
        debugPrint(ref.read(authVM).error);
      }
    },
    child: Text("Anonymous Login"),
  );
}),
```

Now, if we run the application and click the **Anonymous Login** button on the login page, we should be able to log in to the application without providing any email or password.

Next, let's update the login method to use Firebase's email and password-based authentication instead. Open lib/app/view_models/auth.vm.dart and update the login method as follows:

```
// needs verification
Future<bool> login({required String email, required String
password}) async {
  try {
    _auth.signInWithEmailAndPassword(email: email,
      password: password);
    return true;
  }on FirebaseAuthException catch (e) {
    isLoggedIn = false;
    error = e.message ?? e.toString();
    notifyListeners();
    return false;
  }
}
```

Getting the user and updating the logged-in state is again already handled by the auth state change subscription we wrote previously. Once the user is logged in, the SDK keeps track of the session.

Finally, let's see how we register new users in Firebase. In the auth view model, we already have a function called register. Let's update it as follows:

```
Future<bool> register({required String email, required String
password}) async {
  try {
    _auth.createUserWithEmailAndPassword(email: email,
      password: password);
    return true;
  }on FirebaseAuthException catch (e) {
    isLoggedIn = false;
    error = e.message ?? e.toString();
    notifyListeners();
    return false;
  }
}
```

In the preceding code block, we are calling `_auth.createUserWithEmailAndPassword` to create the account. Firebase Authentication also automatically creates a valid session for the user upon signing in. Again, the logged-in state is already handled by the auth state change subscription method we wrote previously. So, we don't need any further changes.

That is all for now for authenticating with Firebase. To learn more about the Firebase authentication service, you can head over to `https://firebase.google.com/docs/auth/flutter/start`. In the next section, we will learn about the Firestore Database service for persisting data.

Using Firestore Database to persist data

Firebase also provides a database service to persist data. There are two databases in Firebase. The first is Realtime Database and the second is Firestore. Firestore is a new, more optimized real-time database, which is why we will be learning about Firestore. Firestore provides flexible, hierarchical data structures. Data is organized in collections and documents. A document can have a complex nested structure as well as subcollections. Data syncs across devices through real-time listeners. To use Firestore, we must first configure it in the console, which we will do in the next subsection.

Configuring Firestore Database

As with Authentication, we have to configure the database in the Firebase console first by following these steps:

1. On the console, click on **Build** | **Firestore Database** from the sidebar.

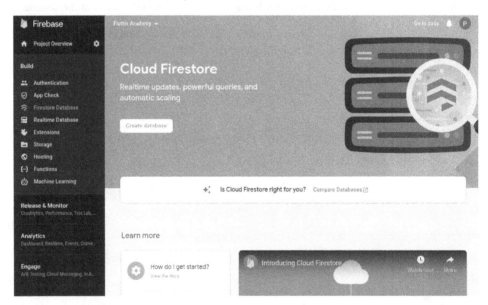

Figure 10.11 – An overview of Firestore

2. On the page that appears, click on the **Create database** button.

3. Select **Start in test mode** so that we are allowed unrestricted access to the database and click the **Next** button:

Figure 10.12 – The Create database dialog

However, when deploying to production, you must remember to secure your data with rules. Learn about the Firestore security rules at `https://firebase.google.com/docs/firestore/security/get-started`.

4. In the next dialog, select the location of your choice. I'll leave it as the default. Then, click on the **Enable** button:

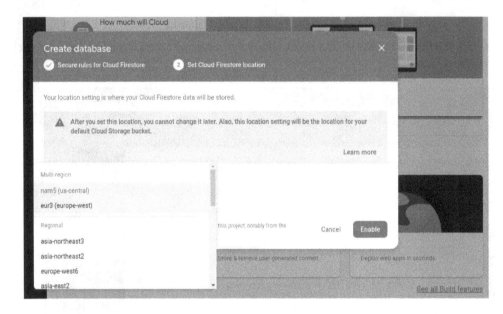

Figure 10.13 – Selecting the Firestore location

Once enabled, you should see the following UI, where you can start creating a collection and adding data. However, we will do that later:

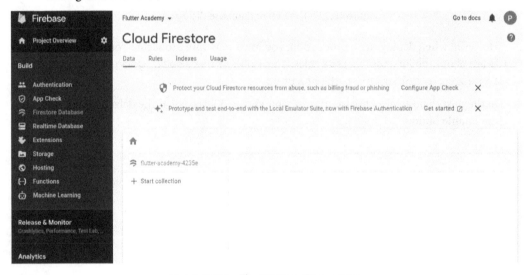

Figure 10.14 – The Firestore Data screen

Now that we have configured Firestore, we will have to configure our application to use Firestore next.

Adding course documents

To add course documents, follow these steps:

1. Open chapter10_start/pubspec.yaml and add Appwrite's Dart SDK:

    ```
    dependencies:
      cloud_firestore: ^3.1.17
    ```

2. Then, run the following command to get the dependencies:

    ```
    flutter pub get
    ```

3. Then, open scripts/load_courses.dart and import the Cloud Firestore SDK first:

    ```
    import 'package:cloud_firestore/cloud_firestore.dart';
    ```

    Next, let's add a function that will create our course collection and add a few courses.

4. Start by making a list of courses that we can add to the database. Prepare a list in the following format:

    ```
    final courses = [
      {
        "title": "Flutter Beginners",
        "description":
            "Awesome course for Flutter beginners to learn
            the basics of Flutter framework",
        "image":
          "https://image-placeholder.com/image.png",
        "status": "published",
        "published_date": FieldValue.serverTimestamp(),
      },
      ...
    ];
    ```

    Add more courses with the same keys as the attributes we have in the database. Make sure to make some course drafts so that we can see how to only show the published courses later in this section.

5. Next, add a function named loadCourses to the same file. Use Firestore to insert those courses into a courses collection:

    ```
    Future loadCourses() async {
      final _db = FirebaseFirestore.instance;
    ```

```
for (final course in courses) {
  _db.collection('courses').add(course);
}
}
```

To load courses, we need to call this function, run the project, and visit /load_courses in the address bar. There, you will find a button – clicking it will add the courses that you can view in the Firestore console:

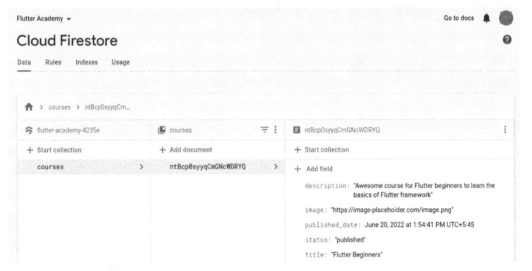

Figure 10.15 – The courses in Firestore

We have successfully inserted courses into Firestore. In the next section, we will list those courses from Firestore in our application.

Listing the courses in our application

To list courses in our application, follow these steps:

1. Open lib/infrastructure/res/course.service.dart, and import and get an instance of the Firestore Database SDK:

    ```
    import 'package:cloud_firestore/cloud_firestore.dart';
    class CourseService {
      final _db = FirebaseFirestore.instance;

      ...

    }
    ```

2. Then, let's update the getCourses function to use Firestore Database to get a list of the courses, as follows:

```
Future<List<Course>> getCourses() async {
  final query = await _db.collection('courses').get();
  return query.docs.map((e) => Course.fromMap(e.id,
    e.data()))).toList();
}
```

Here, we are getting a list of documents in the courses collection and converting it into a course model. Everything else should work as it already is.

Next, we will update our watchlist functionality so that we can save a user's watchlist in Firebase Firestore instead of saving it locally.

Updating the watchlist functionality using Firestore Database

Follow these steps to update the watchlist functionality to use Firestore:

1. Open lib/infrastructure/res/watchlist.service.dart and import the cloud_firestore package first, as we did previously, and update the class with the new properties for the database and collection ID, as follows:

```
import 'package:cloud_firestore/cloud_firestore.dart';
class WatchlistService {
  final collectionId = 'courses';
  final _db = FirebaseFirestore.instance;
  ...
}
```

2. Next, update the addToWatchlist function as follows to use Firestore Database:

```
Future<void> addToWatchlist(String id, String userId)
async {
  await _db.collection(collectionId).doc(id).update({
    "watchlist": FieldValue.arrayUnion([userId])
  });
}
```

Here, we are adding userId to the watchlist field in the course document. We are using FieldValue.arrayUnion because the watchlist field might already have some users in it. We don't want to update the whole field – just add the new value to the existing elements. This will allow us to easily filter the courses that a user is watching based on userId.

3. Next, to remove a course from the watchlist, we can use the following:

```
Future<void> removeFromWatchlist(String id, String
userId) async {
  await _db.collection(collectionId).doc(id).update({
    "watchlist": FieldValue.arrayRemove([userId])
  });
}
```

4. Finally, to get the courses in the watchlist, we get all the courses where the watchlist field contains the user's ID, as follows:

```
Future<List<Course>> getWatchlist(String userId) async {
  final qs = await _db
      .collection(collectionId)
      .where("watchlist", arrayContains: userId)
      .get();
  return qs.docs.map((doc) => Course.fromMap(doc.id,
    doc.data())).toList();
}
```

We also convert the data we get from Firestore to our Course model using the map function of the array.

This concludes database integration. To learn more about Firestore Database, head over to the official documentation for Firestore at https://firebase.google.com/docs/firestore.

Apart from Authentication and Database, Firebase has various other services, such as Analytics, Cloud Functions, and Storage, which help you build robust and rich applications. To learn more about various other services of Firebase, you can always visit the official documentation at https://firebase.google.com/docs.

Summary

In this chapter, we introduced Google's Firebase service. We looked at various services of Firebase. We looked at how Firebase can help us focus on our application and let Firebase handle things in the backend, such as Authentication and Database. We also implemented Firebase Authentication in our application to authenticate our users. Then, we integrated Firestore Database to persist application and user data. In the next chapter, we will learn all about building and deploying our Flutter web application.

Building and Deploying
a Flutter Web Application

In the previous chapter, we learned about Firebase. We learned how to integrate Firebase into our application. We then implemented Firestore Authentication to authenticate our users, and Firestore Database to persist our data.

In this final chapter, we will learn how to build and deploy our Flutter web application. We will talk about the different options for where we can deploy our application as well as learning how to deploy properly. So, let's get started.

By the end of this chapter, you will understand how to build and deploy a Flutter application on the web.

In this chapter, we will cover the following topics:

- Building and deploying your first web app
- Configuring a PWA
- Automating with GitHub Actions

Technical requirements

The technical requirements for this chapter are as follows:

- Flutter version 3.0 or later installed and running
- Visual Studio Code or Android Studio
- Google Chrome
- A GitHub account (get one for free at `https://github.com`)

As this chapter is about deployment, you can just copy the final code from *Chapter 10, Firebase Integration*.

Building and deploying your first web app

Let's start with the most important topic, building and deploying Flutter web applications. In this section, we will first build our application and then deploy it to Netlify. We will mostly use the command line for this. First, open the command line and navigate to the folder containing your Flutter web project. Then, run the following command:

```
flutter build web
```

This creates a release build of the web application for your Flutter project that you can find in <project_root>/build/web. The release build is an optimized build of the Flutter application that is ready to be deployed in production. The release build will have the following structure:

```
/build/web
  assets
    AssetManifest.json
    FontManifest.json
    NOTICES
    fonts
      MaterialIcons-Regular.ttf
      <other font files>
    <image files>
  index.html
  main.dart.js
  main.dart.js.map
```

It might contain more assets or fonts based on your project. During the build, you can also choose the web renderer, which we discussed in *Chapter 2, Creating Your First Web App*:

```
flutter build web -web-renderer canvaskit
```

The web renderer can be `auto`, which is the default, `canvaskit`, or `html`.

Once the build is generated, we can easily deploy it to any static hosting service, including Netlify. To deploy to Netlify, log in to your Netlify account and, on the dashboard, under **Sites**, click on **Add new site**, and select **Deploy manually**:

Figure 11.1 – Netlify dashboard

On the next screen, you need to drag and drop all the files inside `<project_root>/build/web`:

Figure 11.2 – New manual deployment

Once the upload is complete, you should see the following:

Figure 11.3 – Deployed site

Click on the automatically generated link for your newly generated website and you should be able to see it deployed. Now that your site is deployed, you can set up a custom domain or configure further settings by going to **Site settings**. Further configuration of a Netlify deployment is beyond the scope of this book; however, you should be able to find lots of documentation about it.

In this section, we built our Flutter web application and then deployed it using Netlify. That's how simple it is. In the next section, we will look into configuring a **progressive web app** (**PWA**).

Configuring a PWA

In this section, we will understand what PWAs are and how they can be useful. Let's first understand what PWAs are in the next subsection.

Introduction to PWAs

PWA is not an official or formal term. It was first used by Google to refer to web applications that are flexible and adaptable. PWAs are web applications that take advantage of both web and native app features. They are implemented using various technologies and standard patterns. PWAs tend to be faster and more reliable as they use features provided by modern browsers that allow them to access native platform features and work offline. Generally speaking, PWAs are applications written using web technologies that look and behave like web applications; however, by using modern browser technologies, they offer the functionalities of native applications. For instance, they can work offline, send push notifications, and be installed as an application.

PWAs are important for various reasons. First, they are web applications that can be easily accessed via URLs. This means they are indexed by search engines, available while searching, and linkable and widely accessible over the internet. Second, they can utilize native functionalities, just like mobile applications, and can be installed like an application so that users can easily access them from their device by tapping on an icon. PWAs also have good offline support, so they can work without internet access even though they are web pages. These qualities make PWAs very useful. Now that we know what PWAs are and why they are useful, next we will learn about PWAs and Flutter.

Customizing Flutter PWAs

In *Chapter 1, Getting Started with Flutter on the Web*, we noted that Flutter is a great tool for building PWAs. In this chapter, we will look at how we can build PWAs with Flutter. The cool thing about Flutter web apps is that they are already PWAs; we don't have to do much. Flutter web apps, using their plugin system, can already access native functionalities. So, in this section, we will customize the name, icons, and other details of our PWA.

Open web/manifest.json. This is the file where we provide details such as the name, icons, and other configurations. Change the name to Flutter Academy and the short name to FA. Change the description to Flutter academy is an online learning platform where you can learn Flutter from absolute basics. For icons, there are two sizes, 192x192 and 512x512. To change the icons, override the icon files in the web/icons folder with your own icons. That's all there is to it: our PWA is ready to build. To build the PWA, run the following command:

```
flutter build web
```

This command will build the Flutter PWA. The web application will be available in the build/web folder. The files inside this folder comprise the whole application, which we can host and serve using any web server.

We now know what PWAs are and how to configure them. In the next section, we will learn how to automate the build and deployment of our Flutter application using GitHub Actions.

Automating with GitHub Actions

GitHub Pages is another place where we can host static sites by creating a GitHub repository. In order for this to work, all you need is a GitHub account and a repository for your project. We will be using GitHub Actions to automatically build and deploy our Flutter web application to GitHub Pages. First, we will start by creating a GitHub action that will build and deploy our application.

Writing a GitHub action for deployment

We define the configuration for a GitHub action in a `.yml` file inside the `.github/workflows` directory in our repository. Let's create `deploy.yml` inside `.github/workflows` in our repository and add the following configuration:

```yaml
name: Deploy to GitHub Pages
on:
  push:
    branches: [ main ]
jobs:
  build:
    runs-on: ubuntu-latest
    steps:
      - uses: actions/checkout@v2
      - uses: subosito/flutter-action@v1
        with:
          channel: "stable"
      - run: flutter config --enable-web
      - name: Dependencies
        working-directory: chapter10_final
        run: flutter pub get
      - uses: bluefireteam/flutter-gh-pages@v7
        with:
          workingDir: chapter10_final
          baseHref: /Taking-Flutter-to-Web/
```

For this to work, there's one more change we need to make to our app. Open `web/index.html` and replace the following code:

```html
<base href="/">
```

It needs to be replaced with this:

```html
<base href="$FLUTTER_BASE_HREF">
```

This change will allow Flutter to add a dynamic base, `href`. A proper base, `href`, is required for our Flutter application to work, as our application will be deployed in the repository subpath. However, if you later assign a custom domain to your app, this needs to be switched back to the root path (`/`).

The preceding deployment config that we created tells GitHub to run the deployment action whenever there's a change in the `main` branch of the repository. This action will run on an Ubuntu VM using a combination of different action modules to first check out the directory, then install the Flutter dependencies, and finally, run the action to deploy to GitHub Pages. This creates a new `gh-pages` branch and deploys the built page to that branch. You can learn more about the `flutter-gh-pages` GitHub action at their official repository at `https://github.com/bluefireteam/flutter-gh-pages`. Now that our application is built and pushed to GitHub, we need to enable GitHub Pages.

Enabling GitHub Pages

Follow these steps to enable GitHub Pages:

1. First, in your GitHub repository, go to **Settings** | **Pages**:

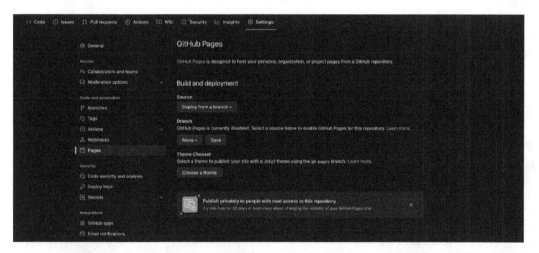

Figure 11.4 – GitHub Pages settings

2. Set **Source** to **Deploy from a branch**, select **gh-pages** for **Branch**, choose to deploy from root, and then click on **Save**, as shown here:

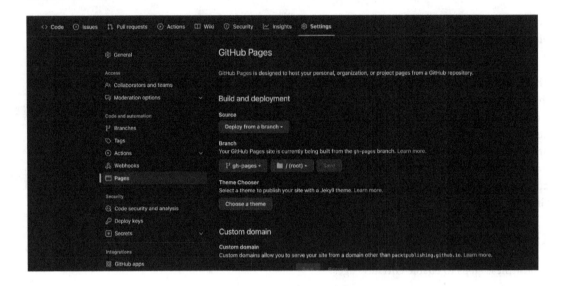

Figure 11.5 – Select the gh-pages branch and save

3. Finally, click on **Choose a theme** and choose any theme. Without choosing the theme, we will not be able to access the deployment properly. However, as we will be deploying our own custom web app built with Flutter, the theme chosen here doesn't matter. After that, you should see the URL for your deployment in the settings:

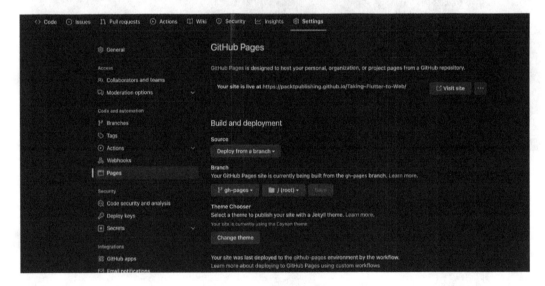

Figure 11.6 – Site complete with URL

If you now visit the site using the URL provided in your settings, you should see that your page has been deployed. Every time you make a change to your application and push it to the `main` branch, within a few minutes, the action should run and deploy the new version automatically.

Summary

In this final chapter, we learned how to build and deploy our Flutter web application so that we can share it with our users. We also learned how we can customize the PWA version of our application by updating the manifest file. Finally, we learned how to use GitHub Actions to automate the build and deploy process so that every time we update our code, the application is automatically built and deployed to GitHub Pages.

Well, that was awesome. Thank you for sticking with me to the end of the book. I hope you enjoyed it. In this book, we started by telling you what Flutter for the web is, why it is essential, and how to get started. We then moved on and created our first web application using Flutter. We then learned about and implemented the responsive and adaptive design in our application. We proceeded further by learning how Flutter for Web works under the hood. Then we learned about routes and navigation to build a multipage application. After that, we learned about architecture and file and folder organization to create a manageable and scalable code base, including MVVM architecture implementation. We then learned about data persistence and state management. We also learned how to integrate with platforms such as Appwrite and Firebase to add advanced capabilities to our application using services such as database, authentication, and others. Finally, we concluded this chapter by learning how to build and deploy our application in production. This book should provide you with the knowledge and tools to build a Flutter application from scratch and finally deploy it to production. And I believe this skill will be precious in the industry.

Index

Other Books You May Enjoy

If you enjoyed this book, you may be interested in these other books by Packt:

Cross-Platform UIs with Flutter

Ryan Edge, Alberto Miola

ISBN: 978-1-80181-049-4

- Create responsive and attractive UIs for any device
- Get to grips with caching and widget trees and learn some framework performance tips
- Manage state using Flutter's InheritedWidget system
- Orchestrate the app flow with Navigator 1.0 and 2.0
- Explore the Material and Cupertino built-in themes
- Breathe life into your apps with animations
- Improve code quality with golden tests, CI setup, and linter rules

A Frontend Web Developer's Guide to Testing

Eran Kinsbruner

ISBN: 978-1-80323-831-9

- Choose the ideal tool or combination of tools for testing your app
- Continuously monitor the market and ensure that your developers are using the right tools
- Advance test automation for your web app with sophisticated capabilities
- Measure both code coverage and test coverage to assess your web application quality
- Measure the success and maturity of web application quality
- Understand the trade-offs in tool selection and the associated risks
- Build Cypress, Selenium, Playwright, and Puppeteer projects from scratch
- Explore low-code testing tools for web apps

Packt is searching for authors like you

If you're interested in becoming an author for Packt, please visit authors.packtpub.com and apply today. We have worked with thousands of developers and tech professionals, just like you, to help them share their insight with the global tech community. You can make a general application, apply for a specific hot topic that we are recruiting an author for, or submit your own idea.

Hi!

I am Damodar Lohani, author of *Taking Flutter to the Web*. I really hope you enjoyed reading this book and found it useful for increasing your productivity and efficiency in Flutter.

It would really help me (and other potential readers!) if you could leave a review on Amazon sharing your thoughts on *Taking Flutter to the Web*.

Go to the link below or scan the QR code to leave your review:

`https://packt.link/r/1801817715`

Your review will help us to understand what's worked well in this book, and what could be improved upon for future editions, so it really is appreciated.

Best Wishes,

Damodar Lohani

www.ingramcontent.com/pod-product-compliance
Lightning Source LLC
Chambersburg PA
CBHW060600060326
40690CB00017B/3772